# Sustainable Dairy Production

Edited by

**Peter de Jong**
Principal Scientist, Processing & Safety
NIZO Food Research BV
The Netherlands

**WILEY-BLACKWELL**

A John Wiley & Sons, Ltd., Publication

This edition first published 2013 © 2013 by John Wiley & Sons, Ltd

Wiley-Blackwell is an imprint of John Wiley & Sons, formed by the merger of Wiley's global Scientific, Technical and Medical business with Blackwell Publishing.

*Registered Office*
John Wiley & Sons, Ltd, The Atrium, Southern Gate, Chichester, West Sussex, PO19 8SQ, UK

*Editorial Offices*
9600 Garsington Road, Oxford, OX4 2DQ, UK
The Atrium, Southern Gate, Chichester, West Sussex, PO19 8SQ, UK
2121 State Avenue, Ames, Iowa 50014-8300, USA

For details of our global editorial offices, for customer services and for information about how to apply for permission to reuse the copyright material in this book please see our website at www.wiley.com/wiley-blackwell.

The right of the authors to be identified as the authors of this work has been asserted in accordance with the UK Copyright, Designs and Patents Act 1988.

*Library of Congress Cataloging-in-Publication Data*

Sustainable dairy production / edited by Peter de Jong, Nizo Food Research Bv, Kernhemseweg 2, 6718 Zb Ede, NL-6718 Z, Netherlands.
    pages   cm
  Includes bibliographical references and index.
  ISBN 978-0-470-65584-9 (hardback)
1.  Dairy engineering.   2.  Dairy products.   3.  Sustainable engineering.
I.  Jong, Peter de, 1965–
  SF247.S86 2013
  636.2'142–dc23
                                                        2012039453

A catalogue record for this book is available from the British Library.

Wiley also publishes its books in a variety of electronic formats. Some content that appears in print may not be available in electronic books.

Cover image: © iStockphoto.com/FrankvandenBergh
Cover design by Meaden Creative

Set in 11/13pt Palatino by SPi Publisher Services, Pondicherry, India
Printed and bound in Malaysia by Vivar Printing Sdn Bhd

1   2013

*For my ancestors in dairy and their children*

# Contents

# Preface

After finishing my chemical engineering studies I chose to go into the dairy sector. For me it had friendly associations, bringing to mind green fields and grazing cows, rather than the smoking chimneys of petrochemical installations. These days, various pressure groups are calling on people not to eat cheese anymore in order to save our planet. Of course this attitude is far too simplistic. As discussed in the introduction to this book, from a nutrient density point of view, dairy products may be among the most sustainable food products. However, as I learned in my chemical training, there's no smoke without fire.

Although it is expected that in the coming years large numbers of new consumers will create new business opportunities for dairy companies, there are still some issues. In 2010 the economic research department of Rabobank concluded that the current global food system is on an unsustainable track, which poses a threat to long-term global food security. In addition, the dairy sector needs to be transformed in order to secure a long-term food supply.

This book aims to provide the dairy and related industries with inspiration for sustainable solutions for today and tomorrow. These solutions are needed because the dairy sector is and will remain essential to feed our children.

*Peter de Jong*
January 2013

# Contributors

**Peter de Jong BSc PhD**
Principal Scientist, Processing & Safety
NIZO Food Research BV
PO Box 20
6710 BA Ede
The Netherlands

**Wil A.M. Duivenvoorden BSc**
Director of Business Development, Industry & Energy
Royal Haskoning
George Hintzenweg 85
3068 AX Rotterdam
The Netherlands

**Pierre Gerber PhD**
Senior Policy Officer, Livestock and the Environment
Animal Production and Health Division (AGA)
Food and Agriculture Organization of the United Nations
Viale delle Terme di Caracalla
00153 Rome
Italy

**Grietsje Hoekstra MSc**
Project Manager, Caring Dairy
CONO Kaasmakers
Rijperweg 20
1464 MB Westbeemster
The Netherlands

**Roel Jongeneel MSc PhD**
Head of Unit Agricultural Policy Group
LEI/Agricultural Economics Institute
Wageningen UR
Alexanderveld 5
2585 DB
The Hague
The Netherlands

**Corine Kroft MSc**
Officer, Corporate Social Responsibility
A-ware Food Group
Handelsweg 5
3899 AA Lopik
The Netherlands

**Erika Mink MSc**
Director, Environment Affairs
Tetra Pak International
81A rue de la loi
B-1040 Brussels
Belgium

**Carolyn Opio MSc**
Natural Resources Officer
Natural Resources Department (NR)
Food and Agriculture Organization of the United Nations
Viale delle Terme di Caracalla
00153 Rome
Italy

**Jaap Petraeus MSc**
Manager, Corporate Environment & Sustainability
Royal FrieslandCampina
P.O. Box 1551
3800 BN, Amersfoort
The Netherlands

**Maartje N. Sevenster MSc PhD**
Sevenster Environmental Consultancy
Isaacs ACT 2607
Australia

**Louis Slangen MSc**
Associate Professor
Agricultural Economics and Rural Policy Group
Wageningen University
Hollandseweg 1
6706 KN Wageningen
The Netherlands

**Arjan J. van Asselt MSc**
Group Leader, Sustainable Processing
NIZO Food Research BV
PO Box 20
6710 BA Ede
The Netherlands

**Klaas Jan van Calker MSc PhD**
Independent sustainability consultant
Sustainability4U
Nijburgsestraat 35a
6668 AX Randwijk
The Netherlands

**Theun V. Vellinga MSc PhD**
Senior researcher, Livestock Production Systems
Wageningen UR Livestock Research
P.O. Box 65
8200 AD Lelystad
The Netherlands

**Michael G. Weeks BE PhD**
Process Performance Manager
Dairy Innovation Australia Ltd
Private Bag 16
Werribee
Victoria 3030
Australia

# 1

# Introduction

Peter de Jong

NIZO Food Research BV, Ede, The Netherlands

**Abstract:** Sustainability is sure to be a major topic in the dairy industry in the years to come. More needs to be invested in new technologies and production chains that result in lower energy consumption and more effective use of milk sources. The reason is not primarily because society seeks a lower carbon footprint for dairy products, but simply because of the scarcity of raw materials. It has been shown that dairy products have a relatively high nutrient density, but also a high environmental impact. It is therefore worthwhile to explore the possibilities for a (much) more sustainable dairy chain, which will help with food scarcity problems in the near future. This book is intended to inspire all those who share this aim. Various points of views are discussed: dairy business and marketing, environmental impact, farmers and supply chains.

**Keywords:** communication, nutrient density, resource scarcity, sustainability hype

*Sustainable Dairy Production*, First Edition. Edited by Peter de Jong.
© 2013 John Wiley & Sons, Ltd. Published 2013 by John Wiley & Sons, Ltd.

## 1.1    Sustainability and the dairy industry: hype or trend?

Some trends cannot be ignored. The global population is steadily growing and an increasing number of mouths need wholesome food in order to stay alive, including our children. It is clear that this situation demands timely and dedicated action from politicians and captains of industry to tackle future resource scarcity. However, such trends are often accompanied by intensive publicity or hype which overestimates the impact of the trend, be it the increasing population or the future shortage of energy resources and drinking water. This hype makes it difficult to define a clear road map for the future. Should we, for example, introduce carbon footprint labelling on food products, or should we be investing heavily in renewable energy, or should we do everything together at the same time?

Up till now, politicians have tried to clarify the impact of population growth and limited resources through detailed analysis. In 2006 the European Commission published a report on the relative impact of products on the environment throughout their life cycle. It was concluded that food and drink are responsible for 20 to 30% of the environmental impact, in which meat and dairy products are most important (European Commission, 2006). A number of studies followed, including from the United States (US Dairy, 2010). National governments have taken the step of setting targets for food companies to reduce energy use and to reduce greenhouse gas emissions. Companies are increasingly obliged to report their use of resources and to agree on reduction targets over a period of 10 years or longer (Government of Australia, 2008; Agentschap NL, 2010).

Since about 2000, a number of food companies have included green annual reports as an addendum to their conventional main annual reports. Such reports were used to communicate companies' good intentions to government, nongovernmental organisations and interested consumers. The main achievements in improving energy efficiency were obtained through closing less-efficient factories and upscaling production (Ramírez-Ramírez, 2006). Nowadays, sustainability has become one of the major messages communicated by a food company and is to be found on the home page of their websites. Companies and branch organisations are transparent about their sustainability goals although it is not entirely clear how sound these goals are and how and when they will be achieved. Here are some examples of food companies' sustainability goals:

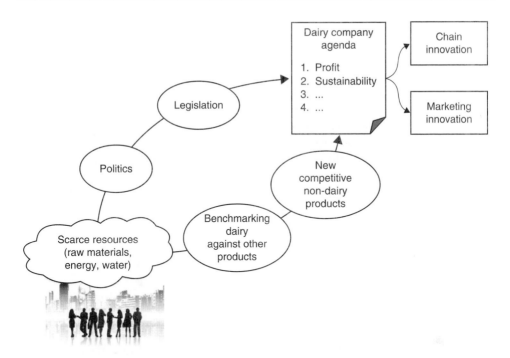

**Figure 1.1    Scarcity means that sustainability must be on the agenda.**

- 'Halve environmental footprint of our products and source 100% of agricultural raw materials sustainably.' (Unilever, 2011)[1]

- 'CO$_2$ neutral in 2020 and global number one through partnerships.' (Danone, 2010)[2]

- 'In 2020 40% of energy used on dairy farms from renewable source, 30% reduction of water use, 50% recycled packaging material.' (Dairy UK, 2008)[3]

In conclusion, it is a clear trend that sustainability will be on the agenda of the worldwide dairy sector for the coming decades. As shown by the simplified scheme in Figure 1.1, scarcity of resources demands more sustainable dairy production.

[1]  www.sustainableenergyforall.org/actions-commitments/ commitments/single/unilever-sustainable-living-plan-reducing-environmental-impact.
[2]  Danone Sustainability Report 2011, www.danone.com.
[3]  www.dairyuk.org/environmental/milk-roadmap.

## 1.2    Quantifying the issue: measuring footprints

Manufacturers of food products are looking for quantitative measures to control the environmental impact of their products. This is not an easy task. A common measure is the carbon footprint, the equivalent of carbon dioxide ($CO_2$-eq) emission per product quantity. Although the calculation in itself is rather simple, the number of factors influencing the carbon footprint is enormous. In the case of dairy, a number of these factors are not known or differ from farm to farm and even from cow to cow. For example, in the United States a recent study showed that farm management, farm size, farm location and forage level accounted for an almost 50% variation in the final footprint (Rotz et al., 2010). This is probably the reason for the large variation in the reported carbon footprint of pasteurised milk. Table 1.1 lists some reported carbon footprints of milk.

This all stresses the need for standardisation and generalisation. The International Dairy Federation, for example, has published a common approach to quantify the carbon footprint for dairy (IDF, 2010). This is a first step towards a standardised measure of carbon footprint for the dairy sector, addressing conversion factors and allocation factors of co-products during manufacturing.

**Table 1.1    Reported carbon footprints of milk (cradle to consumption). Boundary conditions vary between references.**

| Reference | kg $CO_2$-eq per kg milk | Comments |
|---|---|---|
| Carlsson-Kanyama et al., 2003 | 0.3–0.5 | 5 MJ/kg |
| Sevenster & de Jong, 2008 | 0.9–1.8 | Based on literature search in 10 countries (Europe, US, Canada, Oceania) |
| IDF, 2009 | 1.2 | |
| Smedman et al., 2010 | 1.0 | Sweden |
| FAO report: Gerber et al., 2010 | 1.3–7.5 | North America–Africa, global average: 2.4 |
| US Dairy, 2011 | 1.8–2.5 | United States, average 2.1 |

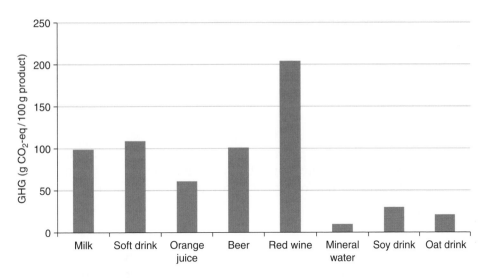

**Figure 1.2   Carbon footprint of several beverages expressed as greenhouse gas (CO$_2$-eq) per 100 grams of product. Adapted from Smedman et al. (2010).**

## 1.3   Communication: telling the whole story

In the discussion of the sustainability of dairy products, several benchmarks have been produced. For example, the carbon footprint of milk has been compared with a number of other beverages such as fruit juices, water and soya milk, as shown in Figure 1.2. Mineral water turns out to be by far the most sustainable choice. However, this is definitely not the whole story! If someone was to consume mineral water only, they would probably die within a month. This leads to the conclusion that labelling of food products with a high focus only on their carbon footprint can be misleading.

Smedman et al. (2010) made a first attempt to relate the climate impact of food products to their nutrient density. They defined a so-called Nutrient Density to Climate Impact (NDCI) index:

$$NDCI = \frac{\text{nutrient density}\left(\text{kcal, protein, vitamins, minerals,}\dots\right)}{\text{greenhouse gas emission}\left(CO_2\text{-eq}\right)}$$

Assuming that a human being needs a certain amount of nutrients to stay alive, the environmental impact is minimal when a person eats food with a high NDCI value. When NDCI is calculated for the products in Figure 1.2, another choice than mineral water becomes sustainable (Figure 1.3). Unsurprisingly,

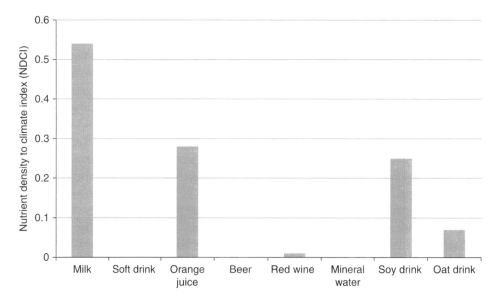

**Figure 1.3   Nutrient density of beverages related to climate impact. Adapted from Smedman et al. (2010).**

milk has the highest NDCI value. In other words, people who drink milk to obtain their daily intake of nutrients and calories have a low impact on greenhouse gases. From this point of view, dairy products are already among the top sustainable food products.

## 1.4   Structure of this book

This book gives an overview of the main aspects of a sustainable dairy production chain. It follows more or less the approach a dairy company can apply to upgrade their degree of sustainability in terms of energy use, carbon footprint and water footprint. At the end some examples of real-life company approaches and new production design concepts are presented.

After this introduction, Chapter 2, 'Greenhouse gas emissions from global dairy production', mainly based on a report from the Food and Agriculture Organization in 2010, examines the subject from a political and social perspective. In Chapter 3 the main tool to assess the degree of sustainability of a product or process is addressed: 'Life cycle assessment'. It explains how to quantify the carbon footprint of a product and which aspects have to be taken into account. The sustainability of the dairy chain is very much based on the farm. This aspect is considered in Chapter 4, 'Sustainability and resilience of the dairy sector in a changing world: a farm economic and EU perspective'. The production of dairy products from raw milk has a long history. Process operations have not changed that much. Chapter 5, 'Dairy processing', outlines some possibilities for increasing the sustainability of dairy processing, and an outlook on breakthrough technologies for the next steps in processing is presented. Besides the farm and the way of processing, packaging has a major impact on the carbon footprint of dairy products, addressed in Chapter 6, 'The role of packaging in a sustainable dairy chain'. Next, two business cases are described: Chapter 7, 'The business case for sustainable dairy products' and Chapter 8, 'A case study of marketing sustainability'. This shows that sustainability is not only a political and social issue but also an opportunity to generate new business and profit. In Chapter 9, 'Cradle to Cradle for innovations in the dairy industry', some thoughts are presented on new concepts to make dairy products 100% sustainable with no negative environmental impact. The book concludes with a vision, 'The future of dairy production and sustainability'.

## References

Agentschap NL (2010) *Long-Term Agreement on energy efficiency in the Netherlands LTA3: Results of 2009.* Document 2MJAA1002. The Hague: Ministry of Agriculture, Nature and Food Quality.

Carlsson-Kanyama, A., Pipping Ekström, M., Shanahan, H. (2003) Food and life cycle energy inputs: consequences of diet and ways to increase efficiency. *Ecological Economics*, 44: 293–307.

European Commission (2006) *Environmental Impact of Products (EIPRO). Analysis of the life cycle environmental impacts related to the final consumption of the EU-25.* Document EUR 22284 EN. Institute for Prospective Technological Studies.

Gerber, P., Vellinga, T., Opio, C., Henderson, B., Steinfeld, H. (2010) *Greenhouse gas emissions from the dairy sector: a life cycle assessment.* FAO report. Rome: Food and Agriculture Organization.

Government of Australia (2008) *National greenhouse and energy reporting guidelines.* Canberra: Department of Climate Change.

International Dairy Federation (2009) Environmental/ecological impact of the dairy sector: literature review on dairy products for an inventory of key issues list of environmental initiatives and influences on the dairy sector. *Bulletin of the International Dairy Federation,* no. 436. Brussels: IDF.

International Dairy Federation (2010) A common carbon footprint approach for dairy: the IDF guide to standard lifecycle assessment methodology for the dairy sector. *Bulletin of the International Dairy Federation*, no. 445. Brussels: IDF.

Ramírez-Ramírez, C.A. (2006) Monitoring energy efficiency in the food industry. Thesis, University of Utrecht, ISBN 9073958989.

Rotz, C.A., Montes, F., Chianese, D.S. (2010) The carbon footprint of dairy production systems through partial life cycle assessment. *Journal of Dairy Science*, 93: 1266–1282.

Sevenster, M., de Jong, F. (2008) *A sustainable dairy sector: global, regional and life cycle facts and figures on greenhouse-gas emissions.* Publication no. 08.778948. Delft: CE Delft.

Smedman, A., Lindmark-Månsson, H., Drewnowski, A., Modin Edman, A.K. (2010) Nutrient density of beverages in relation to climate change. *Food & Nutrition Research*, 54: 5170–5177.

US Dairy (2010) *US Dairy Sustainability Commitment: progress report.* Innovation Centre for US Dairy (www.usdairy.com/sustainability).

# 2

# Greenhouse gas emissions from global dairy production

Theun V. Vellinga,[1] Pierre Gerber[2] and Carolyn Opio[2]

[1] Wageningen UR Livestock Research, Lelystad, The Netherlands
[2] Food and Agriculture Organization of the UN, Rome, Italy

**Abstract:** There is an urgent need to understand what approaches are most effective in reducing greenhouse gas emissions and where to target such efforts in global dairy production. This chapter takes a food chain approach to the estimation of GHG emissions from the dairy cattle sector, assessing emissions from the production of inputs into the production process, emissions related to dairy production, land use change, milk transport and processing.

**Keywords:** assessment, dairy production, greenhouse gas emissions, land use, processing, transport

## 2.1  Introduction

In 2006, the Food and Agriculture Organization published *Livestock's Long Shadow: Environmental Issues and Options* (Steinfeld et al., 2006) which provided the first-ever global estimates of the livestock sector's contribution to GHG emissions,

*Sustainable Dairy Production*, First Edition. Edited by Peter de Jong.
© 2013 John Wiley & Sons, Ltd. Published 2013 by John Wiley & Sons, Ltd.

reckoned to contribute about 18% of total anthropogenic greenhouse gas emissions.

In the wake of the current global climate crisis, it has become increasingly clear that there is an urgent need not only to better understand the magnitude of the livestock sector's contribution to overall greenhouse gas emissions but also to perceive which approaches are most effective in reducing GHG emissions and where to target reduction efforts. Therefore it is necessary to develop a methodology for re-examining global livestock food chain emissions based on a life cycle assessment (LCA) approach. This chapter addresses three broad objectives: first, it aims to disaggregate the initial estimates of the livestock sector's contribution and to assess the dairy sector's contribution to GHG emissions; second, to analyse the effect of intensification on GHG emissions; and third, to identify the major GHG hotspots along the dairy food chain.

The assessment takes a food chain approach in estimating GHG emissions from the dairy cattle sector, assessing emissions from the production of inputs into the production process, emissions related to dairy production, land use change (deforestation related to soybean production), and milk transport (farm to dairy and from processor to retailer) and processing. Given its global nature and the complexity of dairy systems, the assessment relies on several hypotheses and generalisations. The resulting uncertainties were estimated in order to compute a confidence interval.

In this assessment, post-farm-gate emissions are related to a kilogram of milk equivalent at the farm gate and are not estimated for each processed dairy product. Emissions related to processing, production of packaging material and transport for the various dairy products are presented on the basis of an average kilogram of milk at the farm gate, i.e. emissions taking place after the farm gate are attributed back to the milk leaving the farm.

While this chapter deals solely with GHG emissions, it is important to highlight the importance of assessing other sustainability and environmental indicators such as resource use and other non-GHG emission impacts in order to assess the environmental sustainability of production systems. More work needs to be done to examine the trade-offs and synergies with other social and environmental goals, notably impacts on water, air and biodiversity.

## 2.2    Methods for calculating emissions

With increasing concerns over GHG emissions, the need has emerged to assess the total emissions associated with a product. For livestock commodities, life cycle assessment (LCA) of emission intensity can effectively support the identification of strategies to meet the dual challenge of food security and climate change mitigation. This is substantiated by the rapidly growing body of literature underpinning the relevance of LCA to assessing livestock emissions and mitigation options (Cederberg & Mattsson, 2000; de Boer, 2003; Casey & Holden, 2005; Steinfeld et al., 2006; Dalgaard et al., 2007; Garnett, 2007; Thomassen et al., 2008; de Vries & de Boer, 2009; Kool et al., 2009; FAO, 2010). The main strength of LCA lies in its ability to provide a holistic assessment of production processes. The approach also provides a framework to identify effective strategies to reduce environmental burdens, avoiding those that simply shift environmental problems from one phase of the life cycle to another.

Although the principles of LCA are well defined (ISO, 2006), previous studies vary considerably in their level of detail, their definition of system boundaries, the emission factors applied, and other technical aspects such as the allocation techniques and functional units employed. The analysis presented here uses the method, model and database developed by the FAO (2010).

The system boundary is defined by GHG emissions associated with milk production from 'cradle to retail'. Figure 2.1 illustrates the limits of the system studied and includes: (i) emissions associated with feed production, transportation and processing; (ii) emissions associated with fertiliser production, transport and application; (iii) emissions associated with livestock and related manure management systems; (iv) emissions associated with energy use for milking and cooling, lighting and ventilation, agricultural operations (fertiliser and manure application, field operations) and buildings and equipment; (v) emissions related to land use change; and (vi) emissions related to processing and transport of product from farm to processing point and to retailer.

However, excluded from the analysis are emissions related to land use under constant management practices, production of capital goods (buildings and equipment, roads, etc.), and the production of cleaning agents, antibiotics and pharmaceuticals.

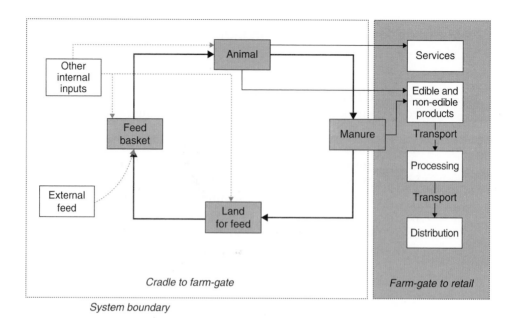

Figure 2.1    **System boundary as defined for this assessment.**

In this assessment, the functional unit used to report GHG emissions was defined as 1 kg of $CO_2$ equivalents per kg of fat and protein corrected milk (FPCM), at the farm gate. FPCM was determined as introduced in Equation 2.1.

$$FPCM(kg) = M(kg) * \{0.337 + 0.116 * FC(\%) + 0.06 * PC(\%)\} \quad [2.1]$$

where M is the mass of raw milk (kg), FC is the fat content (%) and PC is the protein content. All milk was converted to FPCM with 4.0% fat and 3.3% protein.

Milk production is a multifunctional process that generates more than one functional output, for example, milk and meat (from culled cows and surplus calves). Consequently, a part of the emissions from dairy production has to be allocated to the meat as well as the milk. A number of options are available (ISO, 2006); however, for the present analysis, an allocation criterion based on protein content was used to allocate emissions between the two products. While economic allocation has been the most commonly applied allocation criterion (Cederberg & Stadig, 2003; Casey & Holden, 2005; Thomassen et al., 2008), we choose to allocate emissions based on the protein content because the method reflects the primary function of the dairy sector, which is to provide humans with edible protein. The approach also facilitates comparison with other foods and can be applied in situations where markets are absent. The allocation on the basis of protein production, protein and energy, and economic value yielded similar outcomes for milk (FAO, 2010).

The global warming potential for a 100-year period (GWP100) was used to determine the contribution to the GHG effect. Based on the IPCC's *Fourth Assessment Report* (IPCC, 2007), the characterisation factors of GWP100, which corresponds to the $CO_2$ equivalent emission for a period of 100 years, were 1, 25 and 298 for $CO_2$, $CH_4$ and $N_2O$, respectively.

Emissions from the dairy sector were estimated using a combination of Tier 1 and Tier 2 level values with all calculations based on the IPCC guidelines (IPCC, 2006).

The purpose of the assessment was to estimate GHG emissions at global, regional and farming system level. Consequently, a farming system typology was adapted to provide a framework for examining GHG emission from different dairy farming systems. This typology is based on the widely used classification principles set out by Seré and Steinfeld (1996), namely, the feed base and the agro-ecological conditions of production systems.

### 2.2.1  Model description

Based on previous models (Oenema et al., 2005; Schils et al., 2007a; del Prado & Scholefield, 2008), a simplified livestock management model that integrates all major subsystems of a dairy farm was developed to estimate GHG emissions from dairy production systems. The model consists of four submodules: a herd demography module, a feed basket module, an emissions module and an allocation module. Two main methodological innovations have been introduced. A central feature of the model consists of the development of a herd module that computes the 'dairy related stock' (the animal box in Figure 2.1), consisting of the animals required to maintain a population of milking cows and surplus calves that will be fattened for meat production. The second component is a feed basket module that links locally available feed resources with animal numbers and productivity. These two modules permit the computation of information required for the analysis which is not available from statistical databases or literature. In addition, they also ensure coherence between production parameters such as reproduction and herd size, feed intake and milk yields. The conceptual framework of the model is shown in Figure 2.1.

The description of the model provided here is restricted to its salient features; a detailed description of the model was published by FAO (2010).

### 2.2.2   Database and data sources

The input data was obtained from different sources: from statistical databases (FAOSTAT, IFPRI), reports (from CGIAR research institutes, national inventory reports submitted to UNFCCC, etc.), peer-reviewed journals, experts in various fields, as well as specific literature on the sector. The huge data requirements of the LCA have required the development of a global database on herd dynamics (fertility, death and growth rates, etc.), feed (composition, utilisation, quality), manure management systems, crop yield and land use, and so on. To preserve and manage spatial heterogeneity, both at the level of data management and at the level of calculation, the Geographical Information System technique was used in the assessment to create the database and develop the calculation model. Detailed data sources and the main features of the data are described by FAO (2010).

## 2.3   Total emissions of the dairy sector

### 2.3.1   Global overview

The amount of milk produced globally in 2007 was about 553 million tonnes (FAOSTAT, 2009). The amount of meat produced from slaughtered dairy cows and breeding bulls slaughtered after their production period is calculated to reach 10 million tonnes. This meat production is a biologically inevitable co-product of dairy production. The calculated meat production from surplus calves generated by milked cows but not needed for replacement of milked cows and breeding bulls, and thus fattened for beef production, amounts to about 24 million tonnes.

The total meat production related to the global dairy herd is thus calculated at 34 million tonnes, or 57% of global total cattle meat production (60 million tonnes in 2007; FAOSTAT, 2009) and almost 13% of global total meat production, from cattle, sheep, goats, buffaloes, pigs and poultry (269 million tonnes in 2007; FAOSTAT, 2009).

The GHG emissions from the dairy herd, including emissions from deforestation and milk processing, were calculated at 1,989 million tonnes $CO_2$-eq (± 26%), of which 1,328 million tonnes (± 26%] are attributed to milk, 151 million tonnes (± 26%) to meat production from culled and slaughtered animals and 490 million tonnes (± 26%) to meat production from fattened animals (Table 2.1).

Milk and meat production from the dairy herd (comprising milking cows, replacement calves, surplus calves and culled animals) plus the processing of dairy products, production of packaging and transport activities thus contribute 4.1% (± 26%) of the total GHG anthropogenic emissions estimated at 49 giga-tonnes (IPCC, 2007). Milk production, processing and transport alone contribute 2.7% (± 26%) of the total anthropogenic GHG emissions (Table 2.1).

Average global emissions per kg of milk and kg of meat from culled dairy cows and bulls and kg of meat from surplus calves are 2.4 kg of $CO_2$-eq, 15.6 kg $CO_2$-eq and 20.21 kg $CO_2$-eq (± 26%), respectively.

### 2.3.2   *Intensification of dairy production and regional trends*

The global average intensity of milk and meat production shows a wide variation over regions and livestock systems (FAO, 2010).

The mix of technical and management changes undergone by the sector are often designated by the term 'intensification'. Although the term strictly refers to an input/input ratio (e.g. a labour-intensive system is one that use relatively high labour inputs versus capital and land inputs), its common use captures a wide range of trends. Such structural trends have occurred particularly in monogastric food chains, but also for dairy and, to a lesser extent, beef production (Steinfeld et al., 2006, 2010). These trends include:

- Specialisation, increasing scale, and geographical concentration of production; these are ongoing trends wherever the sector is rapidly growing. They are generally driven by economies of scale, economies of scope and transport costs.

- Longer food chains, generally driven by increasing competition for land and the concentration of consumers in urban centres, far from sites of production. Transport is required to bridge the geographical distances between feed and livestock production sites and consumption areas. In addition,

**Table 2.1  GHG gas emissions for main edible goods and overall sector – global averages.**

| Commodities | Total milk and meat production (million tonnes) | GHG emissions (million tonnes CO$_2$-eq)* | GHG emissions (kg CO$_2$-eq per kg of product)* | Contribution to total anthropogenic emissions in 2007 (%)* |
|---|---|---|---|---|
| Milk: production, processing and transport | 553 | 1328 | 2.38 | 2.7 |
| Meat from slaughtered dairy cows and bulls (carcass weight) | 10 | 151 | 15.60 | 0.3 |
| Meat from fattened surplus calves (carcass weight) | 24 | 490 | 20.21 | 1.0 |

*± 26%.

historically decreasing transport costs have allowed the relocation of production and processing activities in the food chains in order to minimise production costs.

- Increased animal productivity and feed conversion ratio, driven by the economic optimisation of production, are achieved by a wide range of technologies, including feeding, genetics, animal health and housing. The shift towards monogastric species, and poultry in particular, has further improved the sector's feed efficiency.

- Pasture intensification through irrigation, fertiliser use and improved grass species; this trend is generally driven by land prices.

Structural changes have a significant influence on the environmental impacts of the sector and also on the options and costs associated with mitigating such impacts. The shift away from traditional mixed and extensive systems has probably had an overall positive effect in improving land and water use efficiency, with negative effects on water pollution, energy consumption and genetic erosion. Nevertheless, the magnitude of the environmental issues has to be attributed to the sector's growth rather than to structural changes per se. Furthermore, traditional and mixed systems would probably not have been able to meet the surging urban demand in many developing countries, not only in terms of volume but also in terms of sanitary and other quality standards. Production intensification is thus seen as an unavoidable trend (Steinfeld et al., 2006, 2010).

The milk yield per animal can be considered as a proxy for these trends in intensification. In general, increases in animal productivity are indeed achieved through the interaction of a variety of factors such as improved nutrition, optimal management, reproduction or genetics; hence, we argue that the choice of milk yield as a measure of intensity is valid as it captures the effects of the above factors in the improved productivity per animal.

There is a significant relationship between milk production per cow and total GHG emissions per kilogram of milk (Figure 2.2). Emissions steeply decrease as productivity increases up to 2,000 kg FPCM per cow per year: from 12 kg $CO_2$-eq/kg FPCM to about 3 kg $CO_2$-eq/kg FPCM. We then observe a slower reduction of emissions as productivity increases to about 6,000 kg FPCM per cow per year. Above this

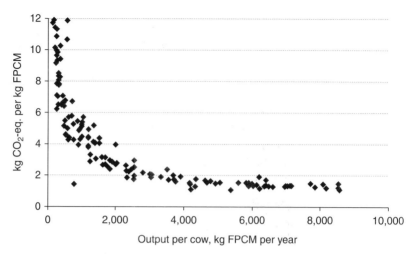

**Figure 2.2   Relationship between total greenhouse gas emissions and output per cow. Each dot represents a country.**

value, emissions stabilise between 1.6 and 1.8 kg $CO_2$-eq/kg FPCM. The general shape of the curve is defined by two factors. First, the fraction of maintenance requirements in the emissions per kg of milk decreases with increasing production per cow. Second, the digestibility of the animals' rations improves from low to high production levels, caused by better roughages and increasing use of high-quality by-products and concentrates. Although the graph suggests a continuous relationship, in fact three classes of dairy systems can be distinguished and, related to them, the production conditions that define the level of greenhouse gas emissions. The very low productive systems rely on feed with a low energy content and a strong seasonal variation in feed availability. Often animals suffer from malnutrition during a number of months of the year. These systems function at a level of survival in a situation of poverty; market access is not the first goal of these subsistence farmers. The second group consists of many smallholders in the developing countries, with varying levels of market access. These farming systems are often multifunctional, optimising the system as a whole, based on spreading of risks. They have limited levels of inputs, but strive to increase productivity. The third group are specialised dairy farms, optimising milk production per cow. Although these farms are specialised, the level of productivity still shows a wide range due to the variation in agro-ecological and socio-economic conditions and, hence, a wide variation in input levels.

The asymptotic relation between emissions and yield is given by Equation 2.2. The model explains 89.3% of the variance and gives an asymptotic value of emissions of 1.37 kg $CO_2$-eq/kg FPCM.

$$\ln(GHG) = 0.3174 + 2.3837 * \left(0.9993320^{YIELD}\right) \qquad [2.2]$$

where GHG expresses the total $CO_2$, $N_2O$ and $CH_4$ emissions, in $CO_2$-eq, and YIELD is the output per cow, in kg FPCM per year.

There is a significant relationship between $CH_4$ and $N_2O$ emissions per kg FPCM and milk production per animal ($R^2=0.97$ and $R^2=0.82$, respectively; Figure 2.3). Trends are similar to those described above for total GHG emissions. Carbon dioxide emissions are relatively constant, with more variation in intensity in low milk yields than in higher yields. However, no significant correlation was established between carbon dioxide emissions and milk yields.

The fractions of the three GHG emissions per kg of FPCM in relation to milk production per cow are shown in Figure 2.4. The figure reveals the increasing share of $CO_2$ in total emissions with productivity gains ($R^2=0.73$). The trend can be explained by the increased input of energy via fertilisers, machinery and the extra purchased feed used in the higher input systems. They carry a $CO_2$ cost of production. The shares of $CH_4$ and $N_2O$ emissions decrease as the share of $CO_2$ increases, but no significant correlation could be established for these gases.

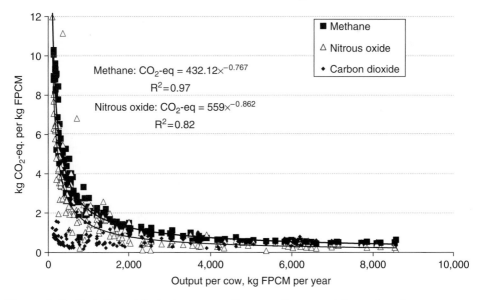

**Figure 2.3    Relationship between methane, nitrous oxide and carbon dioxide emissions and output per cow. Each dot represents a country.**

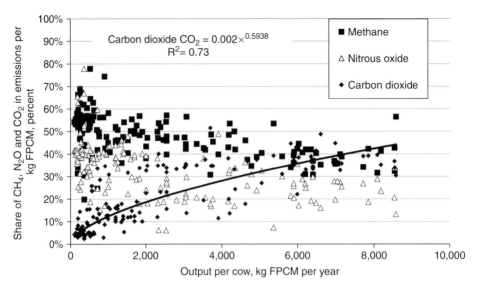

**Figure 2.4    Fraction of methane, nitrous oxide and carbon dioxide in total GHG emissions, in relation to output per cow. Each dot represents a country.**

### 2.3.3    Post-farm-gate emissions

The proportion of milk processed in dairy plants and the mix of commodities produced varies by region. In industrialised countries, 95 to 100% of the milk produced is transported to the dairy plant for processing. The remainder is generally processed on-farm into cheese, butter and yogurt; a limited amount of raw milk is sold fresh. In most developing countries, however, transport infrastructure and market are limited: usually, all milk is sold locally or processed to butter and cheese by the milk-producing household.

The International Dairy Federation provides data for a number of countries, mostly the industrialised countries, and representing 74% of global raw milk production. In these countries, 85% of all raw milk enters dairy plants for processing (see Table 2.2; IDF, 2009).

Six major dairy products can be considered: fresh and fermented milk, cream (and related butter), cheese, whey and milk powder. The processing chains and global average partitioning of milk are shown in Figure 2.5.

Significant regional differences exist in the relative importance of dairy products. For instance, cheese is quite important in the EU-27 and North America; on the other hand, in New Zealand and to some extent Australia, milk powder takes precedence (Table 2.3).

**Table 2.2    Percentage of raw milk transported to dairy plant for processing in regions included in IDF reports.**

| Region | Share of raw milk sent to dairy plant (%) |
| --- | --- |
| North America | 96 |
| South America | 82 |
| Asia | 62 |
| EU-27 | 89 |
| Other European countries | 78 |
| Oceania | 100 |

*Source*: IDF (2009).

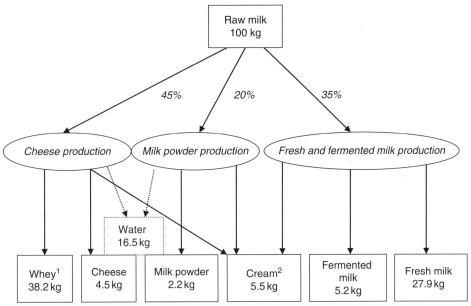

1. Whey is sold as feed and as whey powder.
2. Cream is sold as such or processed into butter.

**Figure 2.5    Milk processing chains and related mass partition: a global average.**

Regional variations are considerable, related to differences in end-products, energy sourcing and energy efficiency. For example, emissions are relatively high in Australia and India due to a high percentage of coal use in energy production (Figure 2.6).

**Table 2.3 Milk processing: regional variations in mix of end-products.**

| Region/ country | Fresh milk (% of raw milk) | Fermented milk (% of raw milk) | Cheese (% of raw milk) | Condensed milk (% of raw milk) | Milk powder (% of raw milk) |
| --- | --- | --- | --- | --- | --- |
| EU-27 | 25 | 8 | 52 | 3 | 12 |
| Australia | 26 | no data | 33 | no data | 34 |
| New Zealand | no data | no data | 19 | no data | 52 |
| Canada | 37 | 4 | 45 | 2 | 11 |
| USA | 31 | 2 | 51 | 1 | 10 |
| Average for countries and regions above | 26 | 6 | 51 | 3 | 14 |

*Source:* IDF (2009).

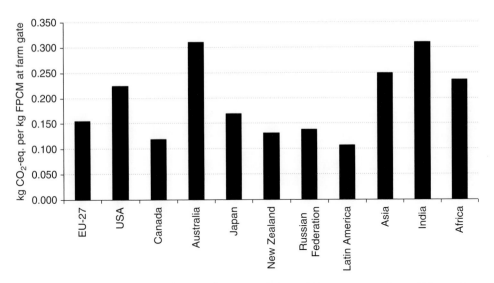

**Figure 2.6   Calculated GHG emissions between farm gate and retail from the processing of 1 kg of raw milk in selected countries and regions.**

Averaged over all milk produced in a country, post-farm-gate emissions range between 0.06 and 0.23 kg $CO_2$-eq per kg of milk. This is the combination of the fraction of processed milk in a country and the emission intensity per kg of processed milk, as shown in Figure 2.6. We can conclude that the contribution of milk processing to the total carbon footprint of milk is limited.

### 2.3.4   Sensitivity, uncertainty and validation

#### Sensitivity analysis

Many factors affect the efficiency of dairy production systems and GHG emissions per kg of milk and meat. Limited data availability has necessitated several simplifications and assumptions. Uncertainty regarding emission factors (IPCC, 2006) is an additional source of potential error. A sensitivity analysis showed that changes in feed digestibility, combined with improved animal performance and manure management, strongly affect the level of GHG emissions. The results for herd and feed parameters are shown in Figure 2.7.

#### Accuracy

The sensitivity analysis has shown that the emissions per kg of milk and meat are mostly affected by digestibility, milk yield per

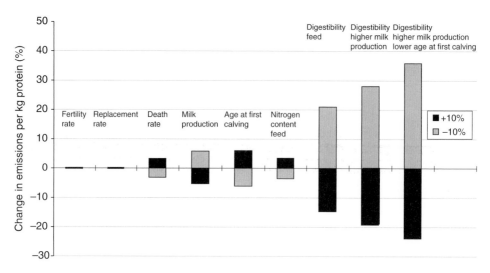

**Figure 2.7   Sensitivity analysis: effect of a 10% change in key parameters on GHG emissions per kg of animal protein from a dairy system (including fattening calves).**

cow and manure management. The subsequent uncertainty analysis, with random variation of input parameters and emissions factors, showed that the confidence interval of the estimated emissions ranges between plus and minus 26% of the calculated average figure of 2.4 kg $CO_2$-equivalents per kg of FPCM.

### Validation

The slaughtered animals and total meat production figures calculated using the herd demography module were compared to FAO statistics (FAOSTAT, 2009) and found to be very similar for all countries, except for a few countries where live animals are traded in large numbers.

Calculated GHG emissions were also compared to previous studies based on similar methodologies. Methane emissions per animal from this assessment are comparable to figures obtained by Schils et al. (2007b) and Cederberg et al. (2009) in OECD countries (ranging from 110 to 130 kg methane per cow per year) and by Herrero et al. (2008) in Africa, ranging between 21 and 40 kg methane per livestock unit per year.

Emissions per kg of milk compare well with prior LCA studies for dairy production (Table 2.4) (Foster et al., 2007; Vergé et al., 2007; Blonk et al., 2008; Capper et al., 2008; Herrero et al., 2008;

**Table 2.4  Results from prior life cycle assessment studies of dairy production from cradle to farm gate.**

| Reference | Country/region/system | $CO_2$-eq per kg of milk | $CO_2$-eq per kg of meat | Remarks |
|---|---|---|---|---|
| Basset-Mens et al., 2009 | New Zealand | 0.65–0.75 | — | High maize yields, special emission factors |
| Foster et al., 2007 | United Kingdom | 1.14 | | |
| Vergé et al., 2007 | Canada | 1.0 | | |
| Blonk et al., 2008 | Netherlands | 1.2 | 8.9 | |
| Sevenster & DeJong, 2007 | Annex 1 countries | 0.75–1.65 | — | Based on national Inventory Reports/ UNFCCC data |
| Thomassen et al., 2008 | Netherlands | 1.5–1.6 | — | |
| Capper et al., 2009 | USA | 1.35 | — | |
| Cederberg et al., 2009 | Sweden | 1.00 | 19.8 | Allocation milk/meat 85/15, meat including beef cattle |

Sevenster & de Jong, 2008; Thomassen et al., 2008; Basset-Mens et al., 2009; Cederberg et al., 2009). Some of the results from prior analyses are lower than those presented in this chapter, which in part is caused by discrepancies in emission factors (e.g. Basset-Mens et al., 2007; Cederberg et al., 2009) or allocation technique (Cederberg et al., 2009). The decision to use the standard emissions factors of the IPCC at Tier 2 level may also result in discrepancies when compared to studies that utilise country-specific emissions factors.

## 2.4   Discussion

### 2.4.1   Contribution to climate change

The assessment presented here only looks into the GHG emissions of the dairy sector. It is obvious that GHG emissions are one aspect only of the environmental sustainability of the sector, which also includes issues such as water resource preservation, biodiversity erosion and air pollution. Furthermore, environmental performance is only one of the criteria against which the sustainability of production systems is achieved, others being social issues, public health and profitability.

The results and conclusions of this chapter need to be understood in this context, and analysed considering the synergies and trade-offs existing among environmental objectives and between environmental and other objectives. Whereas, for example, we estimated that intensification of production is coupled with a reduction of GHG emissions per unit of output, the implications regarding eutrophication of water resources, biodiversity conservation or social fabric may well go in the opposite direction.

### 2.4.2   Efficiency and potential for mitigation

The combined production of milk and meat is particularly efficient in achieving low GHG emissions per unit of product. The fundamental biological reason for this is the fact that milk is a 'non-extractive' product which is harvested without any reduction of the generating biomass (stock). The emissions associated with growing a calf into an adult animal, and maintaining the animal until it is slaughtered, are thus attributed among the production of both beef and milk; whereas they are entirely attributed to beef in specialised beef systems. Still, there is scope for emissions reduction. In production, the main mitigation avenues are to limit methane and nitrous oxide emissions.

In intensive systems, enteric methane emissions per kg of milk are relatively low, compared to extensive systems, leaving relatively little opportunity for improvement. In contrast, the fraction of methane coming from manure storage is relatively high (15 to 20%, compared to less than 5% in the extensive systems of the world's arid and humid zones). Anaerobic digestion of manure to produce biogas is already a proven technique which thus has a significant potential. In the extensive systems of the arid and humid zones, marginal improvements of feed digestibility could achieve significant reductions in methane emissions per kg of milk, through a direct reduction of emissions and through the improvement of milk yields (Kristjanson & Zerbini, 1999).

The high contribution of nitrous oxide to the emissions of extensive systems in the arid and humid regions is mainly caused by the deposition of dung and urine in pastures, due to the long grazing time for the animals, and through the use of dry-lot type manure storage. Where feasible, changing manure management in these regions could be an effective way to reduce emissions.

Sequestering carbon by increasing soil organic matter content in grassland is a further way to offset emissions. Natural grasslands represent about 70% of the world's agricultural lands. Improving grazing land management is estimated to have the highest mitigation potential among all possible agricultural mitigation sources, over 1.5 billion tonnes $CO_2$-eq per year (IPCC, 2007). The restoration of degraded grasslands through erosion control, revegetation and improved fertility also has significant potential to increase soil carbon sequestration rates. This can also generate additional ecosystem services relating to water quality and biodiversity management, and improve the productivity and resilience of livestock enterprises.

Overviews of mitigation options have been published and provide useful information (Schils et al., 2006; Steinfeld et al., 2006; Smith et al., 2008). Special attention should be paid to trade-offs and displacement of emissions among production stages (Wassenaar et al., 2007; van Groenigen et al., 2008). Analysis shows that the effectiveness of mitigation options depends on the specificities of the livestock systems.

With regard to post-farm activities, mitigation options include choosing packaging material for which production and disposal emit limited amounts of GHG, as well as choosing energy sources with a low emission level.

## References

Basset-Mens, C., Ledgard, S., Boyes, M. (2009) Eco-efficiency of inten-sification scenarios for milk production in New Zealand. *Ecological Economics*, 68(6): 1615–1625.

Blonk, H., Kool, A., Luske, B. (2008) *Milieueffecten van Nederlandse con-sumptie van eiwitrijke producten. Gevolgen van vervanging van dierlijke eiwitten anno 2008*. Gouda: Blonk Milieu Advies.

Capper, J. L., Castaneda-Gutierrez, E., Cady, R. A., Bauman, D. E. (2008) The environmental impact of recombinant bovine somatotro-pin (rbST) use in dairy production. *Proceedings of the National Academy of Sciences of the USA*, 105(28): 9668–9673.

Casey, J.W., Holden, N.M. (2005) Analysis of greenhouse gas emissions from the average Irish milk production system. *Agricultural Systems*, 86: 97–114.

Cederberg, C., Mattsson, B. (2000) Life cycle assessment of milk pro-duction: a comparison of conventional and organic farming. *Journal of Cleaner Production*, 8: 49–60.

Cederberg, C., Stadig, M. (2003) System expansion and allocation in life cycle assessment of milk and beef production. *International Journal of Life Cycle Assessment*, 8: 350–356.

Cederberg, C., Sonesson, U., Henriksson, M., Sund, V., Davis, J. (2009) *Greenhouse gas emissions from Swedish production of meat, milk and eggs 1990 and 2005*. Göteborg: SIK, Swedish Institute for Food and Biotechnology.

Dalgaard, R., Halberg, N., Hermansen, J.E. (2007) *Danish Pork Production, and Environmental Assessment*. University of Aarhus, Aarhus, Denmark.

de Boer, I.J.M. (2003) Environmental impact assessment of conven-tional and organic milk production. *Livestock Production Science* 80: 69–77.

del Prado, A., Scholefield, D. (2008). Use of SIMSDAIRY modelling framework system to compare the scope on the sustainability of a dairy farm of animal and plant genetic-based improvements with management-based changes. *Journal of Agricultural Science*, 146: 195–211.

de Vries, M., de Boer, I.J.M. (2009) Comparing environmental impacts for livestock products: a review of life cycle assessments. *Livestock Science* 128: 1–11.

FAO (2010) *Greenhouse gas emissions from the dairy sector: a life cycle assessment*. Food and Agriculture Organization, Rome, Italy.

FAOSTAT (2009) FAO Statistical Database. from Food and Agriculture Organization: http://faostat.fao.org/site/339/default.aspx.

Foster, C., Audsley, E., Williams, A., Webster, S., Dewick, P., Green, K. (2007) *The environmental, social and economic impacts associated with liquid milk consumption in the UK and its production: a review of litera-ture and evidence*. London: Defra.

Garnett, T. (2007) Meat and dairy production and consumption: exploring the livestock sector's contribution to the UK's greenhouse gas emissions and assessing what less greenhouse gas intensive systems of production and consumption might look like. Working paper. Centre for Environnmental Strategy, University of Surrey, Surrey, UK.

Herrero, M., Thornton, P.K., Kruska, R., Reid, R.S. (2008) Systems dynamics and the spatial distribution of methane emissions from African domestic ruminants to 2030. *Agriculture Ecosystems & Environment*, 126(1–2): 122–137.

International Dairy Federation (2009) The world dairy situation 2009. *Bulletin of the International Dairy Federation*, No. 438. Brussels: IDF.

IFPRI (2009) *Agro-maps – mapping of agricultural production systems.* www.ifpri.org/dataset/agro-maps-mapping-agricultural-production-systems.

IPCC (2006) *IPCC Guidelines for National Greenhouse Gas Inventories, Volume 4 Agriculture, Forestry and Other Land Use.* Hayama, Japan: Intergovernmental Panel on Climate Change, IGES.

IPCC (2007) *Climate Change 2007: IPCC Fourth Assessment Report.* Cambridge: Cambridge University Press.

ISO (2006) *Environmental Management – Life Cycle Assessment: requirements and guidelines.* Geneva: International Organization for Standardization.

Kool, A., Blonk, H., Ponsioen, T., Sukkel, W., Vermeer, H., de Vries, J., Hoste, R. (2009) *Carbon footprints of conventional and organic pork: assessment of typical production systems in the Netherlands, Denmark, England and Germany.* Wageningen University, the Netherlands: Blonk Milieuadvies.

Kristjanson, P.M., Zerbini, E. (1999) *Genetic enhancement of sorghum and millet residues fed to ruminants: an ex ante assessment of returns to research.* LRI Impact Assessment Series 3. Nairobi, Kenya: International Livestock Research Institute.

Oenema, O., Wrage, N., Velthof, G.L., van Groenigen, J.W., Dolfing, J., Kuikman, P.J. (2005) Trends in global nitrous oxide emissions from animal production systems. *Nutrient Cycling in Agroecosystems*, 72(1): 51–65.

Schils, R.L.M., Verhagen, A., Aarts, H.F.M., Kuikman, P.J., Sebek, L.B.J. (2006) Effect of improved nitrogen management on greenhouse gas emissions from intensive dairy systems in the Netherlands. *Global Change Biology*, 12(2): 382–391.

Schils, R.L.M., de Haan, M.H.A., Hemmer, J.G.A., van den Pol-van Dasselaar, A., de Boer, J.A., Evers, A.G., et al. (2007a) DairyWise, a whole-farm dairy model. *Journal of Dairy Science*, 90(11): 5334–5346.

Schils, R.L.M., Olesen, J.E., del Prado, A., Soussana, J.F. (2007b) A review of farm level modelling approaches for mitigating greenhouse gas emissions from ruminant livestock systems. *Livestock Science*, 112(3): 240–251.

Seré, C., Steinfeld, H. (1996) *World livestock production systems – Current status*. FAO Animal Production and Health Papers no. 127. Rome: FAO.

Sevenster, M., de Jong, F. (2008) *A sustainable dairy sector: global, regional and life cycle facts and figures on greenhouse-gas emissions*. Publication no. 08.7798.48. Delft: CE Delft.

Smith, P., Martino, D., Cai, Z., Gwary, D., Janzen, H., Kumar, P., et al. (2008) Greenhouse gas mitigation in agriculture. [Review]. *Philosophical Transactions of the Royal Society B-Biological Sciences*, 363(1492): 789–813.

Steinfeld, H., Gerber, P., Wassenaar, T., Castel, V., Rosales, M., & de Haan, C. (2006) *Livestock's long shadow: environmental issues and options*. Rome: FAO.

Steinfeld, H., Mooney, H.A., Schneider, F., Neville, L.E. (Eds.) (2010) *Livestock in a changing landscape: drivers, consequences and responses. Volume 1*. Washington, London: Island Press.

Thomassen, M.A., van Calker, K.J., Smits, M.C.J., Iepema, G.L., de Boer, I.J.M. (2008) Life cycle assessment of conventional and organic milk production in the Netherlands. *Agricultural Systems*, 96(1–3): 95–107.

van Groenigen, J.W., Schils, R.L.M., Velthof, G.L., Kuikman, P.J., Oudendag, D.A., Oenema, O. (2008) Mitigation strategies for greenhouse gas emissions from animal production systems: synergy between measuring and modelling at different scales. *Australian Journal of Experimental Agriculture*, 48(1–2): 46–53.

Vergé, X.P.C., Dyer, J.A., Desjardins, R.L., Worth, D. (2007) Greenhouse gas emissions from the Canadian dairy industry in 2001. Agricultural Systems Special Section: sustainable resource management and policy options for rice ecosystems. *International symposium on sustainable resource management and policy options for rice ecosystems*, 94(3): 683–693.

Wassenaar, T., Gerber, P., Verburg, P.H., Rosales, M., Ibrahim, M., Steinfeld, H. (2007) Projecting land use changes in the Neotropics: the geography of pasture expansion into forest. *Global Environmental Change: Human and Policy Dimensions*, 17(1): 86–104.

# 3

# Life cycle assessment

Maartje N. Sevenster

Sevenster Environmental Consultancy, Isaacs, Australia

**Abstract:** Life cycle assessment is a quantitative method that is used as a decision support tool. It gives insight into which parts of the production chain have high impacts on the environment and what interdependencies exist between different life cycle phases. Recent results and development of a methodology of life cycle assessment for dairy products are presented.

**Keywords:** carbon footprint, emissions, energy use, life cycle assessment, methodology

## 3.1  Introduction

Life cycle assessment (LCA) is a quantitative method that supports so-called life cycle management (LCM) and life cycle thinking (LCT) in both industry strategy and government policy-making. It is first and foremost a decision support tool and its results are in principle valid only in the context of the specific goal of an LCA study.

*Sustainable Dairy Production*, First Edition. Edited by Peter de Jong.

The idea of assessing product life cycles developed in the late 1960s when it was focused primarily on material flows (resource use and waste production). This resource and environmental profile analysis (REPA) then developed in the late 1970s into the full energy cycle analysis, a logical focus after the energy crisis. The development of biofuels in Brazil was already partly based on such energy analysis, much like today's biofuels are judged by their net benefit in terms of greenhouse gas emissions.

In this chapter, LCA as it is today will be discussed. The focus will be on its application to the life cycle of dairy products, from cultivation of feed crops to the processing of milk into final products. First, the constituents of LCA will be described (section 3.2), with a brief discussion of the application of LCA in practice (section 3.3). Then, recent results and methodology development of LCA of dairy products are presented (section 3.4). The chapter ends with a section on the reasons why business and government need to be concerned with life cycle thinking and LCA.

## 3.2   Current life cycle assessment

It was in the early 1990s that the concept of impact assessment was added to life cycle assessment. Until then, the analysis dealt with inventories of materials, emissions and energy use, but there was an increasing demand to translate these parameters into actual impacts on the environment, such as 'climate change' or 'acidification'.

Those two important constituents of LCA, the inventory (LCI) and the impact assessment (LCIA), are still the main building blocks of life cycle studies. The development of impact assessment was primarily a European idea; in the USA, even today, LCA results are often in terms of energy use and emissions (inventory items) rather than their effects. The main development of so-called characterisation factors for different impacts took place at the Institute of Environmental Sciences in Leiden, Netherlands (Gabathuler, 2006). In this methodology, the impacts were expressed as 'potential damage' in a number of environmental impact categories that were important in Dutch policy at the time. Typical categories are climate change, acidification, emissions that are toxic to humans and smog formation.

These so-called 'mid point' categories are still used in many of the current, more developed impact methods. An additional group of impact methods, however, goes one step further and expresses impacts of so-called 'end-point' categories. These are actual damage to human health, to ecosystems and to natural resources.

The advantage of end-point categories is obvious: true damage is what society is concerned about. Also, different impacts on human health may be directly compared, whereas 'smog formation' and 'human toxicity' are two separate quantities. The disadvantage is that the calculation of true damage is more complicated and involves many more uncertainties than the calculation of potential damage in terms of the mid-point categories. Thus, results in terms of end-point categories are considerably less certain.

It is the LCA practitioner's choice whether to use a mid-point or an end-point approach to impacts. Next to the building blocks already discussed, the ISO standards on LCA (ISO 14040: 2006 and 14044: 2006) also distinguish the goal and scope definition and the interpretation phases as essential constituents of an assessment (Figure 3.1). In the goal and scope definition, choices are made concerning system boundaries, allocation (see section 3.4.1), impact categories and other scope parameters, all with reference to the goal of the study. An important choice is whether the assessment is to be consequential or attributional and this very much depends on the questions that need to be answered by the study. The interpretation phase then deals with discussion of uncertainties, crucial parameters

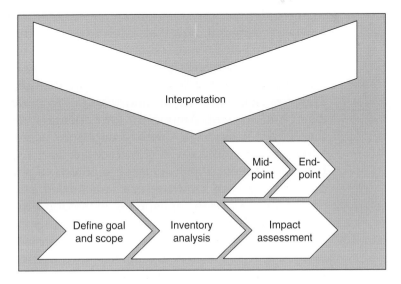

**Figure 3.1   The building blocks of life cycle assessment.**

in the life cycle and general discussion of the results against the background of the predefined goal. A first interpretation round may lead to redefinition of earlier phases (iteration). For instance, if initial assessment points to certain processes that make a large contribution to the overall impacts, one may go back to those processes and try to decrease uncertainty of the related data.

### 3.2.1  Impacts associated with land use

During the first 30 years of the development of LCA, assessment dealt only with tangible inputs and outputs of processes: materials including energy carriers in, emissions and waste out. With the onset of end-point impact assessment in terms of ecosystem damage, however, the issue of land use and related effects began receiving attention. As defined in the *Handbook of Life Cycle Assessment* (Guinée et al., 2002) the damage associated with land use comprises loss of biodiversity, loss of life support function and land competition (loss of land as a resource).

While initially land use was not really regarded as an essential inventory item, in recent years it is emerging as one of the central issues, together with climate change. For agricultural products, such as dairy, this is of course an important development. When considering plain hectares as a measure for land use, grazing systems typically have higher land use per unit than more intensive systems, but may have lower energy requirements and the like (see also section 3.4). From an environmental point of view, however, different types of land use have different effects. Typically, pasture lands or orchards have lower biodiversity impacts than croplands, simply due to the more intensive management of the latter which causes more disturbance to fauna. Therefore, the intensity of land use should preferably also be taken into account. Among the first attempts at incorporating the effects of land use ('land occupation') were the methods known as Eco-indicator 99 (Goedkoop & Spriensma, 1999) and EPS 2000 (Steen, 1999). In those methods, a square metre of different types of land use is characterised by a 'potentially disappeared fraction' or 'species extinction'. In other words, these methods target the loss of biodiversity as the main impact category. As yet, there are no generally accepted measures for those impacts, but several impacts methods offer indicative solutions, among others, the current versions of the two methods mentioned.

## 3.2.2  *Global, regional and local effects*

Different impacts act on different scales. This complicates the interpretation of LCA results. For emission of greenhouse gases, the situation is relatively simple, as climate impacts are global: the damage from emission of 1 kg of carbon dioxide does not depend noticeably on the location of emission. This means that all emissions in a life cycle of a product, no matter where they occur, may be given the same environmental factor.

Emissions of sulfur dioxide, nitrous oxide, ammonia and so on act on regional scales. Impacts primarily occur within some 2,000 km of the emission source. As the impacts depend on the population density in that zone, and on the vulnerability of ecosystems, they may be very different depending on the exact location of the source. A unit emission of ammonia in New Zealand will result in very different damage compared with a unit emission of ammonia in the Netherlands.

Emissions of particles by traffic, or noise and odour nuisance, act on very local scales; impacts can in fact be considered to be zero if there are no people around to experience the hindrance. This is reflected in zoning policies which keep industrial activities separate from housing. In specific air quality studies, the health damage from a unit of emission from traffic in a metropolitan area is found to be orders of magnitude higher than for the same unit of emission in a rural area.

How do the effects of land use rank on this scale? Actual effects may be seen at a regional scale, due to spillover effects on biodiversity in surrounding ecosystems (habitat fragmentation, edge effects). Also, large-scale changes at the surface of the Earth may lead to dramatic changes in regional climate, influencing other agricultural areas as well as ecosystems. Nevertheless, the total impacts of a unit of land use or land change are very much determined at a local level.

Other than for greenhouse gases, effects at a regional or local scale should therefore theoretically not be compared without proper differentiation. As has been shown in a recent study (Wegener Sleeswijk & Heijungs, 2010), emissions of pesticides lead to impacts that may differ by a factor of 10 to a million for different locations around the world. In practice, however, such differentiation is not yet commonplace. In a typical LCA study, budget constraints are tight and there is simply no time for such detail. Standard impact factors are applied regardless of the location of emission. Sensitivity assessment can be used to judge the uncertainty that is thus introduced. Although existing software packages have been designed to differentiate impacts to some extent, application at the required level of differentiation is still far off. It will require a major effort to determine the characterisation factors on coarse- to fine-meshed global grids for a range of impact categories and to adapt LCA software correspondingly.

### 3.2.3 *Water use*

Another topic that is finding its way into life cycle assessment is the use of water and effects associated with it. In the *Handbook of Life Cycle Assessment* (Guinée et al., 2002), only desiccation (desertification) is mentioned as a water-related impact category. However, over-availability of water may also cause environmental problems. The effects of water use are also very much local and may even depend on the time of year (e.g. dry or wet season[1]). Therefore, water 'footprinting' requires very detailed attention. A major difference between approaches is whether or not water evaporation (evapotranspiration) from crops is taken into account. This so-called 'green water' use causes high water footprints, even if no active irrigation ('blue water') takes place (see Hoekstra et al., 2011, for methodology details). The footprinting method of Hoekstra et al. gives an indication of how the hydrological balance in the areas of production may be influenced, but not necessarily of all water use in the life cycle processes. Cooling water (non-consumptive water use) is excluded, as long as the water is returned to the same catchment area from which it was extracted. On the other hand, heat pollution by cooling water may have to be counted ('grey water').

The water footprint including 'green' water of a litre of milk is estimated to be 1,000 litres of water[2] as a global average. Evaporation of water from feed crops is the main constituent of this footprint. A point of discussion is whether this 'gross' green water footprint is a good measure, or whether the 'net' green water use should be used. That is, whether the evapotranspiration of crops should be corrected for the natural evapotranspiration that the local ecosystem would have undergone anyway. To assess the influence on the hydrological balance, the net

---

[1]  Impacts caused by emissions can also vary significantly with weather conditions, but emissions are typically assumed to be constant year round and impacts can thus be averaged over the year. Water use cannot be assumed to be continuous, especially in agriculture.

[2]  www.waterfootprint.org.

evaporation gives more insight, but the current consensus on the footprinting method is to assess the total 'appropriation' of water (see Hoekstra et al., 2011).

## 3.3   Life cycle assessment in application

As already shown in Figure 3.1, LCA consists of four phases, as defined by the ISO standards. Those phases are:

- goal and scope definition;
- inventory;
- impact assessment;
- interpretation.

The phases are not meant to be simply sequential; iteration is an essential part of executing LCA. Table 3.1 gives some more detailed steps involved in the four phases, with a brief description.
   We may distinguish the following principal applications:

- Single product: identify 'hot spots' in the life cycle;
- Comparisons: compare products or product designs (alternatives);
- Footprinting: establish absolute 'impact' within a certain predefined framework, meant to allow comparison between different studies (e.g. carbon footprinting, see Chapter 2);
- Policy support/evaluation: assess effects of structural changes in economy due to policy measures, e.g. stimulating the use of biofuels or changing dietary patterns.

The goal and scope definition and the interpretation may be the most important aspects of LCA. Too often practitioners take a 'one size fits all' approach, for example, always using the same allocation method or system boundary definition regardless of the goal of the study. Once the goal and scope have been well defined, data collection may be painstaking and time consuming but is in principle straightforward. Impact assessment is easy with a wealth of impact models available in LCA software. The complexity of LCA is primarily in the interpretation of the results. Results should not be over-interpreted, that is, used as an answer to questions that do not match the goal of the study. Also, both in

**Table 3.1 Life cycle assessment: constituents.**

| Phase | Topic/terminology | Description |
|---|---|---|
| Goal and scope definition | Goal | Explicit definition of the goal of your particular LCA study |
| | System boundaries | Should be predefined, in context of goal. Example: if studying the life cycle effects of a major change in consumption from meat to cheese, the effects of extra beef from dairy cows as a result of extra dairy production should also be taken into account. |
| | Functional unit | Define actual function that the assessment refers to, e.g. 'kg of milk', 'kg of cheese' or 'kg of protein' |
| | Other choices | Should be predefined, in context of goal. Concerns, e.g., allocation (see discussion in 3.4.1) |
| Inventory analysis | Data quality | According to goal and scope definition. Data quality needs to be high for comparative studies that are disclosed to public, especially if claims are made to be better than competing product. If goal is initial insight into relative contributions of processes, lower data quality may suffice |
| | Foreground/ background data | Foreground data concern, e.g., the emissions of enteric fermentation and energy use in milking machines in a dairy LCA; background data would be the emissions associated with a unit of energy use |

| | Allocation | In some cases, only a fraction of a certain inventory item (e.g. energy) should be allocated to the product. As predefined in scope (see discussion in section 3.4.1) |
|---|---|---|
| Impact assessment | Classification | Assign impact categories to all inventory items, as predefined in scope |
| | Characterisation | Assign impact factors (characterisation factors) to all inventory items according to method predefined in scope |
| | Normalisation | Relate impact results to certain reference (optional, as predefined in scope); often the total emissions of Europe or the world are used for this. The normalised results then show the contribution (in %) of the product life cycle to the total impacts of that geographical region. Easily interpreted as implicit weighting |
| | Grouping | Optional, as predefined in scope; this allows for defining groups of impact results that may have, e.g., higher or lower political relevance. Easily interpreted as implicit weighting |
| | Weighting | Officially part of the impact assessment (ISO, 2006a, 2006b) but in practice interpretative function as weighting allows comparison of results of different impact categories. As weighting always involves value choices it is not allowed in comparative assertions intended to be disclosed to the public (ISO, 2006a, 2006b) |
| Interpretation | | |

a single-product assessment, when establishing the processes in the life cycle with highest environmental impacts, and in a product comparison, the results may be different for different impact categories. What should we conclude when product A has a lower climate change impact than product B, but higher toxic impacts?

The weighting of results of various impact categories can thus be an important aid in the interpretation phase. Weighting is in all cases a value-based choice and therefore contentious. Many LCA practitioners prefer to base their assessment exclusively on natural sciences, as is advocated by the ISO standards. The latter explicitly preclude weighting in the case of publicly reported comparative assertions.

However, as a decision support tool, LCA would not be very useful if the decision-maker was always faced with a set of contradictory results. As LCA is gaining ground as an assessment tool for general policy support and evaluation, this is increasingly the case. Originally, LCA was meant to be a tool for product design. This meant comparing several versions of one product; in this case the alternative options studied would in their basis be very similar. Contradictory results for different impact categories are less likely in this situation. If they occur, product design may simply be re-optimised in such a way that trade-offs disappear and weighting is not necessary.

In LCA for policy development, quite different products may be compared or even significant changes in the overall economy. For instance, the comparison of biofuels with traditional fossil fuels involves two totally different production life cycles and with a target of 10% replacement in 2020, such as proposed in the European Union, the changes cannot be considered marginal. Naturally, the functional unit of the two life cycles is the same, otherwise a comparison would not be possible at all. Apart from that, however, one is dealing with agriculture, fertiliser use and potential deforestation on the one hand, and extraction of oil, emissions of volatile organic compounds and extensive capital goods on the other. In those circumstances, contradictory results are almost inevitable and weighting is essential to move beyond the 'we don't know' stage.

Such weighting may be done in a variety of ways. In essence, three general approaches exist:

- distance to target, e.g. using policy targets or reduction costs;
- damage costs, using economic valuation of damage;
- panel weighting, applied to normalised results.

Basically, weighting is determined by an 'external' factor such as economic damage, political relevance or contribution compared to some predefined reference or simply expert judgement. With all three approaches, the relative importance of different environmental categories may be established. The importance is only valid within the context of the value choice that is made, however. In the damage cost approach, damage to human health typically dominates the results. Using panel weighting, ecosystem damage is more prominent and, in weighting based on policy targets, climate change is dominant in countries with well-defined greenhouse gas reduction policies. To prevent a biased interpretation of results, one may apply several weighting methods to see whether they lead to the same overall conclusions. In the context of policy-making or target evaluation, target-based weighting should be avoided, for obvious reasons.

## 3.4  Life cycle assessment of dairy products

Before looking at LCA studies of dairy, a better idea of what the dairy life cycle looks like may be helpful. Figure 3.2 provides an overview of the major steps in the full cradle-to-grave life cycle of dairy products. At several steps, material inputs are required, as well as energy inputs (not shown graphically). At almost all steps, co-products are generated in processes, starting from the use of oilseeds for meal (feed) and oil (co-product) to the processing of dairy with several co-products. Even the waste treatment of packaging may generate energy as a 'co-product'.

Many LCA studies focus on the 'cradle to farm gate' life cycle or, in other words, on the production of raw milk. A typical functional unit is either a kg of milk as produced or a kg of energy-corrected milk (ECM) or fat- and-protein-corrected milk (FPCM).

For processed dairy products, the production of the raw milk typically makes by far the dominant contribution to overall life cycle impacts. As a product of animal husbandry, the most important environmental impacts are:

- climate change, emissions of methane (enteric fermentation) and $N_2O$ (manure, fertiliser use for feed crops);

- climate change, emissions of $CO_2$ due to land management or land clearance for grazing and feed crops. Carbon uptake in grasslands may or may not be taken into account (see Chapter 2);

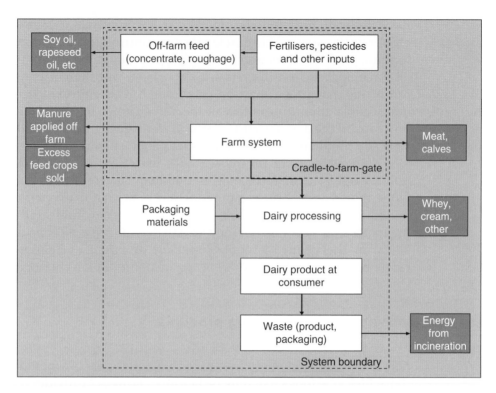

**Figure 3.2    Flowchart for a typical dairy product. Dark grey boxes indicate possible co-products.**

- eutrophication (excess of nitrogen or phosphates) due to manure;

- acidification, e.g. due to formation of ammonia from manure and urine;

- land use, both for grazing and for feed production.

Determining the inputs and outputs for processes in a livestock system is much less straightforward than for many industrial processes. The production of raw milk does not just entail a dairy cow converting feed into milk and emissions, but also requires the maintenance of replacement stock and bulls and it produces calves and meat (see Figure 3.2). Therefore, the entire herd needs to be taken into account in the modelling. Replacement rate is an important variable in determining the life cycle efficiency of a livestock system. Because of the co-production of beef, allocation is also a crucial issue in dairy LCA. This is discussed in section 3.4.1.

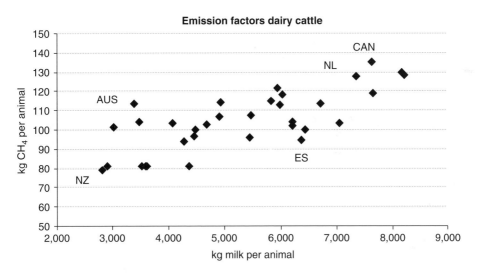

**Figure 3.3    Methane emission from enteric fermentation versus yield per animal. AUS, Australia; CAN, Canada, ES, Spain; NL, Netherlands; NZ, New Zealand.** *Source*: Sevenster & de Jong (2008). Reproduced with permission from CE Delft.

Another crucial variable is the yield of milk per animal. Over the years, the yield has increased enormously in optimised systems such as those in Northern Europe and America.

In Figure 3.3, the Netherlands and Canada stand out as countries with high milk yield. The methane emissions of enteric fermentation per animal are also high in these countries. A trend of increasing emission per animal with increasing yield is apparent. New Zealand is at the other end of the distribution. Australia and Spain are included as outliers: one with fairly low yield but high methane emission per animal, the other with higher yield but relatively low emissions.

In Figure 3.4, the same emissions of enteric fermentation are shown, in terms of kilograms methane per animal as well as kilograms methane per kg of milk produced. From a life cycle perspective, with functional unit one kilogram of product, the latter is the most interesting. The high-yield countries show relatively low emissions per kg of milk produced. The lowest score is for Spain, however, which has the highest milk yields.

Of course, enteric fermentation is just one of the sources of emissions in the life cycle. Initially (at low yields), increasing the yield per animal in fact also leads to overall life cycle environmental efficiency, but at higher yields the benefit of extra milk production may be cancelled by the requirement for extra feed and the production of nitrogen-rich manure. One may suspect that from a life cycle perspective there is actually an

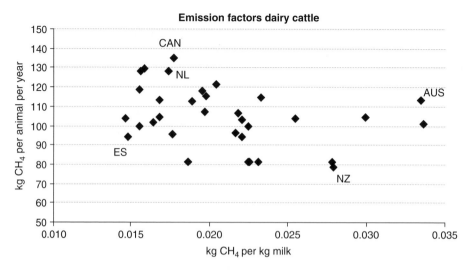

**Figure 3.4     Methane emission from enteric fermentation, per animal and per kg milk. AUS, Australia; CAN, Canada, ES, Spain; NL, Netherlands; NZ, New Zealand.** *Source*: Sevenster & de Jong (2008). Reproduced with permission from CE Delft.

optimal yield range that results in lowest overall impacts per litre of milk produced, although this has as yet not been explicitly established. As we will see further on, the full life cycle of New Zealand milk production is very efficient from an environmental perspective even though the yield per animal is relatively low.

### 3.4.1  Allocation

Allocation refers to the partitioning of accumulated environmental burdens over several co-products that are formed in one and the same process. In dairy LCA, allocation plays an even more important role than usual due to the inextricable co-products of milk and meat (along with other 'products' such as pulling power, leather, offal, etc.). In Figure 3.2, co-products are indicated in dark grey boxes. Each flow that crosses the system boundary requires allocation. When trying to assess the environmental impacts associated with a dairy product, one has to make a decision about how to account for the fact that meat is also produced in the life cycle. This concerns the meat not only of the dairy cow itself but also of some of her calves that are fattened up for beef. In a situation with such inextricably linked co-products, several choices can be made by the LCA practitioner to deal with this.

The first choice is to expand the product system, from containing just the dairy product to including also the associated meat. In that case, all life cycle impacts can be unambiguously counted toward the product system, but the double functionality is not usually attractive. The impacts of say '100 kg cheese plus 2 kg beef' are hard to compare to anything else. A derived method is so-called 'substitution'. In that approach, the impacts of 2 kg beef from a beef-only production system are subtracted from the impacts found via system expansion. This method is disputed for dairy and beef, however, because the quality of beef is not necessarily comparable.

The second choice is to find a physical or science-based mechanism that explains a causal relation between environmental burden and the co-products. For dairy cattle there is an intriguing new development here in using the physical conversion of feed by the animal as a lead. For different feed ingredients it is known what part of the energy content goes toward milk production and what part to growth and maintenance of the body (i.e. meat). An elaborate study by Thoma and Jolliet (2010) in the US shows that the resulting allocation factor can be modelled as a simple function of the ratio of meat (kg) to milk (kg) produced, which is easily determined in a case study. If the ratio of meat to milk is R, the allocation factor is $[1 - 5.7717*R]$ (IDF, 2010). With R of the order of 0.025 for more developed systems, the allocation to milk would be 85.6%.

This provides an obvious mechanism for the allocation toward milk and meat of the impacts associated with feed. As most other impacts on farm are digestion related – enteric fermentation and manure – those may be allocated using the same partition. This approach is applied in the International Dairy Federation's footprint standard (IDF, 2010). Currently, these physical conversion parameters are only known for cattle, not for other dairy animals. Also, variation between cattle types may need further study.

If such a physical relation is not known, as is often the case, further options for allocation are available. Roughly, they are the substitution method and allocation by proxy, in some sense the only true allocation method. The proxy can be mass, in a very simple approximation, or energy or protein content, or economic value. The latter is often used as it reflects the actual economic driving force that dictates production and thus environmental burden in the first place. For example, this is an often preferred method to allocate burdens between oilseed meals and oil (IDF, 2010).

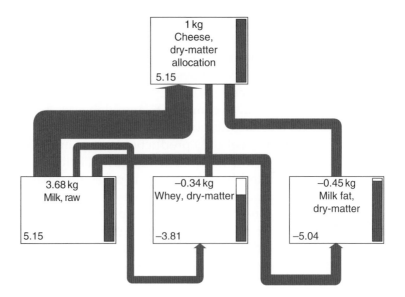

**Figure 3.5   Allocation of impacts toward co-products (dry-matter allocation, fictional example).**

Another complex issue is the allocation to co-products of the dairy processing stage. Even in the simplest case there is cream and standardised milk. One step up and we find cream, whey and cheese. The most complex case can be found in the dairy refinery where a dozen different dry ingredients – lactose, casein, etc. – are produced from a single litre of milk. So how much of the impacts of the production of raw milk should be counted toward the cheese, how much toward the cream and how much toward the whey? The same choices of mass, economic value, etc., are available and used in practice. Simply using mass is not realistic in this case, because then in the refinery plant, most impacts would go toward the water evaporated to yield the products. This obvious non-result immediately leads to the solution of using dry-matter (milk solids) content as a proxy. So cream gets a fairly high allocation per unit volume because it has a high dry-matter content. Whey gets a low allocation per unit volume, but still a fairly high allocation in total due to its high volume.

In Figures 3.5 and 3.6, this dry-matter allocation is compared to economic allocation for a fictional low-fat cheese production step. In both cases, 10 kg of milk is used to make 1 kg of cheese, along with whey and cream as co-products. The difference in net impact allocated to the cheese is a factor of almost two. The total environmental impact is 14 (sum of numbers in lower left-hand corner of

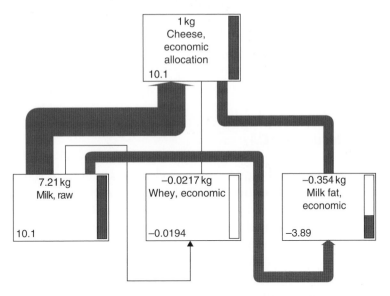

**Figure 3.6   Allocation of impacts toward co-products (economic allocation, fictional example).**

three boxes at bottom of the figures). In the first allocation approach, almost equal amounts of this go toward cheese (5.15 impact units), whey (3.81 impact units) and fat (5.04 impact units). In the second approach, cheese takes two-thirds of the load (10.1 impact units) and virtually nothing goes toward the whey (less than 0.02 impact units). It can also be seen that in the first case (Figure 3.5), of the 10 kg of milk, only 3.68 kg are attributed to the cheese, while in the second case (Figure 3.6) this is 7.21 kg.

The choice of allocation has a large effect on the results, as demonstrated by Figures 3.5 and 3.6. Of course, co-products only take part of the burden when they are going toward useful application. If they go to waste, on the other hand, those waste processes become an extra burden for the remaining product. So, if a cheese factory emits whey straight to surface water or soil, the environmental burden of the cheese is higher than in a theoretical process that yields no whey at all. If the cheese factory sells the whey as pig feed or as a dry ingredient to the food industry, the environmental burden of the cheese is lower than in that theoretical process that yields no whey. In this sense, there is a double benefit of avoiding waste.

As soon as the products are physically separated, the life cycles are separated as well. The energy required for drying whey to powder is not part of the cheese life cycle and should not be counted in the impacts of cheese. If energy requirements are not known to such level of detail for a factory, however, it is possible to apply a generic allocation matrix as developed by Feitz et al. (2007). This matrix gives generic, relative shares of different products in the energy use of dairy processing based on an analysis of a large number of New Zealand milk processing factories.

### 3.4.2  Results of LCA

In the literature, life cycle assessments of dairy production in several countries can be found. Although care should be taken in comparing results of different studies directly, given all the potential differences in scope definition, functional unit, data quality, etc., it is interesting to look at the larger-scale differences between systems.

Based on four different studies, we have data on global warming, acidification, eutrophication, energy use and land use for dairy systems in four countries. Some studies assessed true average production for the entire country, others used a large sample of farms to represent the average production. For three of the countries, conventional and organic systems were explicitly distinguished. The results are compared in Figure 3.7.

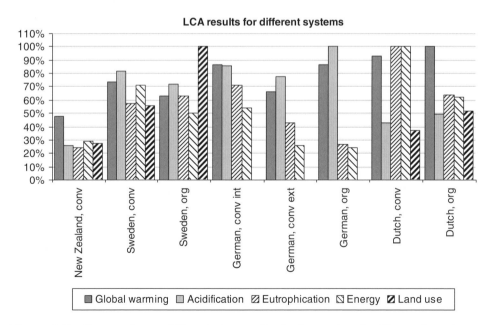

**Figure 3.7   Comparison of literature results for impacts of milk produced under a variety of systems.** *Source*: Bassett-Mens et al. (2005); Thomassen et al. (2008); Cederberg & Mattson (2000).

It is important to note that the different indicators or environmental aspects all point in a different direction when it comes to finding the 'worst' score over the life cycle. Therefore, the overall ranking of systems in terms of life cycle impact cannot be made without a subjective choice of weighting factors (see discussion on weighting in section 3.3).

Interestingly, however, all indicators point to New Zealand as the most efficient system (an equal score to Germany for energy use) in terms of these environmental indicators per kg of milk produced. This is remarkable, because the yield per cow in New Zealand is not very high (see Figure 3.3). Compared to the Dutch systems, with highest scores on three of the five indicators, the yield in kg milk per cow is only a third. The Dutch dairy systems can be considered much more efficient from that point of view, but this does not lead to better 'eco-efficiency' from a life cycle perspective due to the high inputs and outputs of the Dutch system (see Figure 3.3 and discussion there).

In Bassett-Mens et al. (2005) the efficiency of the New Zealand system is explained by the very high yield of grass per hectare in the favourable climatic conditions, the use of clover for nitrogen fixation and the almost 100% grazing system which means manure management does not require transport, storage or other handling. Such a system cannot necessarily be replicated in other regions.

It is impossible to distinguish between the other systems in terms of eco-efficiency in an objective (scientific) manner. Several sets of weighting factors exist to provide extra insight into the relative importance of different environmental aspects (see section 3.3), but these are always value based. Another option would be to translate the results from mid-point to end-point indicators, as discussed in section 3.2. This would reduce the number of results and allow direct comparison between, say, land use and climate change in terms of effect on ecosystem quality. It would also allow distinction between land use for pasture and land use for broad-acre crop production, which is not visible in this mid-point approach that simply counts hectares.

The question is whether it is useful to compare milk life cycle impacts between different systems. To some extent it is, but a more usual goal of LCA would be to improve eco-efficiency within one system. In that case, improvements that do not entail trade-offs between environmental aspects should always be preferred. LCA is an excellent tool for determining whether such trade-offs exist.

## 3.5   Life cycle assessment in strategy and policy

LCA, in particular carbon footprinting (see Chapter 2 and Chapter 7), is in principle the domain of business, as instrument for decision-making, customer communication or even voluntary self-regulation. Large players in retail, such as Walmart and Tesco, have started schemes to move toward a situation in which all suppliers will need to provide information on impacts or footprints of their products. An interesting example from a completely different sector is the green purchasing scheme set up by Prorail, the company responsible for railway infrastructure in the Netherlands. Prorail has developed an elaborate system of assessing the carbon awareness of suppliers, with different well-defined levels. A high level according to this system leads to a higher probability of a contract being awarded. In 2011, this system was adopted by a separate foundation that aims at an even wider application of this method (see www.skao.nl, in Dutch only).

Closer to home, major dairy producers around the globe such as FrieslandCampina, Arla Foods and Fonterra are assessing many of their products and packaging using LCA. Unilever has its own international LCA department, assessing and monitoring brands and products in the company's portfolio. Business itself is thus integrating the entire life cycle of their product(s) in their development and marketing strategies. Life cycle management is becoming a standard component in strategic decisions, partly due to pressure from consumers and NGOs. Products that are clearly visible to the consumer and that have emotional value, such as chocolate and dairy products, have been the first to experience this pressure but also to realise the potential benefits of life cycle management. Detailed knowledge about sourcing of raw materials, ingredients and feed is becoming part of doing 'business as usual'.

In addition, governments are increasingly focusing on the life cycle impacts of economic activities and consumption in general. The European Commission launched its Integrated Product Policy in 2003, which targeted LCA and product design, as well as green procurement, as instruments to stimulate greener products. The Netherlands have been on the front line of life-cycle-based policy, by designing the first policy instrument ever that explicitly included life cycle energy reduction. This memorandum of understanding (or covenant) between government and several industrial sectors, including the dairy processing industry, states that energy reduction in the life cycle of the products of participating industries counts toward the overall energy reduction target. Similarly explicit reduction targets counted over entire life cycles of product groups have been introduced in waste and resource policy. In terms of life cycle effects of overall consumption, dairy has been listed along with meat as one of the focus areas in studies for the European Commission (Tucker et al., 2006; Weidema et al., 2008). In a more recent study for the Dutch government, dairy turned out to contribute approximately 15% to the impacts of average food consumption by Dutch citizens (Sevenster et al., 2010).

We can thus identify a variety of drivers for business to investigate their products' life cycles. Pressure from NGOs and consumer groups has certainly been a factor in the past. Increasingly, however, business itself is taking the lead and targeting cost reduction, added value and risk minimisation. Ignorance of the structure and effects of a product's life cycle may entail serious risks. This may apply mostly to the upstream life cycle, i.e. all the processes that take place before a company's own processes, but can also apply to the downstream life cycle. Sectors involved in the production of base materials may not be well aware of the end-user sectors where their products are ultimately applied. Changes in seemingly unrelated sectors may propagate upstream and unexpectedly influence one's direct market. For dairy products, with a relatively short life cycle and major producers only one step away from the consumer, this is less of an issue. Upstream effects can be an issue, however. With ever increasing demand for limited resources, materials and energy, andland and water are becoming scarce and expensive and prices may be volatile. Arable land is certainly one of the resources that will be subject to increasing pressures. Measuring, and if possible minimising, the amount of those resources needed to produce a unit of product therefore makes perfect sense from the perspective of longer-term business strategy. Being less dependent on limited resources will mean more stability in the future.

## Acknowledgements

During the writing of this chapter the author was a consultant with CE Delft (www.ce.nl).

# References

Basset-Mens, C., Ledgard, S., Carran, A. (2005) *First life cycle assessment of milk production from New Zealand dairy farm systems*. S.l. (N.Z.): AgResearch Limited.

Cederberg, C., Mattson, B. (2000) Life cycle assessment of milk production: a comparison of conventional and organic farming. *Journal of Cleaner Production*, 8: 49–60.

Feitz, A., Lundie, S., Dennien, G., Morain, M., Jones, M. (2007) Generation of an industry-specific physico-chemical allocation matrix. *International Journal of LCA*, 12(2): 109–117.

Gabathuler, H. (2006) The CML story: how environmental sciences entered the debate on LCA. *International Journal of LCA*, 11(Suppl. 1): 127–132.

Goedkoop, M.J., Spriensma, R.S. (1999) *The Eco-indicator 99, Methodology report. A damage oriented LCIA Method*. The Hague: VROM.

Guinée, J.B., Gorrée, M., Heijungs, R., Huppes, G., Kleijn, R., de Koning, A., et al. (2002) *Handbook on life cycle assessment: Operational guide to the ISO standards*. Dordrecht: Kluwer Academic.

Hoekstra, A., Chapagain, A., Aldaya, M., Mekonnen, M. (2011) *Water footprint assessment manual*. London/Washington: Global Footprint Network.

International Dairy Federation (2010) A common carbon footprint approach for dairy. *Bulletin of the International Dairy Federation*, no. 445. Brussels: IDF.

ISO (2006a) ISO 14040: 2006 Environmental management – Life cycle assessment – Principles and framework.

ISO (2006b) ISO 14044: 2006 Environmental management – Life cycle assessment – Requirements and guidelines.

Sevenster, M., de Jong, F. (2008) *A sustainable dairy sector; facts and figures*. Delft: CE Delft.

Sevenster, M. (CE Delft), Blonk, H., van der Flier, S. (Blonk Milieu Advies) (2010) *Environmental impact analysis of food and food losses as high-priority flows in lifecycle-based waste policy* [in Dutch]. Delft, January 2010.

Steen, B. (1999) *A systematic approach to environmental priority strategies in product development (EPS). Version 2000, General system characteristics, CPM report*. Gothenburg, Sweden: Chalmers University of Technology.

Thoma, G., Jolliet, O. (2010) A causal approach to allocation between milk and beef for dairy life cycle analyses. Abstract. LCAX conference, November 2010, Portland, Oregon (www.lcacenter.org/LCAX/index.shtml).

Thomassen, M., van Calker, K., Smits, M., Iepema, G., de Boer, I. (2008) Life cycle assessment of conventional and organic milk production in the Netherlands. *Agricultural Systems*, 96: 95–107.

Tukker, A., Huppes, G., Guinée, J., Heijungs, R., de Koning, A., van Oers, L., et al. (2006) Environmental Impact of Products - analysis of the life cycle environmental impact related to the final consumption of the EU-2. EIPRO.

Wegener Sleeswijk, A., Heijungs, R. (2010) GLOBOX: a spatially differentiated global fate, intake and effect model for toxicity assessment in LCA. *Science of the Total Environment*, 408(14): 2817–2832.

Weidema, B.P., Wesnæs, M., Hermansen, J., Kristensen, T., Halberg, N. (2008) Editors: Peter Eder and Luis Delgado, Environmental Improvement of Products (IMPRO). *Environmental Improvement Potentials of Meat and Dairy Products*. Institute for Prospective Technological Studies.

# 4

# Sustainability and resilience of the dairy sector in a changing world: a farm economic and EU perspective

Roel Jongeneel[1] and Louis Slangen[2]

[1] Wageningen UR, The Hague, The Netherlands
[2] Wageningen UR, Wageningen, The Netherlands

**Abstract:** A framework is discussed in which the resilience and sustainability of the dairy sector is investigated with respect to environment, agronomy, ecology and economics. As indicators of economic sustainability and resilience, attention is paid to profitability, costs of production and competitiveness. The picture generated by the set of indicators that is analysed is that dairy production is only partially sustainable from a profit maximisation perspective. Policies can play an important role in improving the dairy sector's profitability. As compared to the past, income support and environmental sustainability issues should be better integrated, for example by introducing a better targeting of payments made to farmers.

**Keywords:** Europe, farmers, politics, profitability, resilience, subsidies

*Sustainable Dairy Production*, First Edition. Edited by Peter de Jong.
© 2013 John Wiley & Sons, Ltd. Published 2013 by John Wiley & Sons, Ltd.

## 4.1  Introduction

### 4.1.1  Background

The dairy sector remains one the most import sectors within agriculture. It offers people livelihood, status, income and employment. This holds both for developed (e.g. the EU) as well as developing and emerging economies (e.g. India). In developing countries cattle often play a dual role as draught animal and wealth symbol, and are a key source of animal protein in otherwise protein-poor diets.

Table 4.1 provides an overview of world dairy production and the countries that are the main players. World production of raw milk is about 700 million tonnes and is expected to grow by about 2% per annum. Whereas currently, developed countries dominate world dairy production, in the coming years this is going to change. The reason is that the expected growth in production is relatively low in the developed world (in the OECD[1] area, only 0.7% per annum), but high in the developed world (more than 3% per annum). As Table 4.1 shows, the EU-27 is the world's biggest dairy producer, followed by India. The US is third in terms of production. However, importance in production does not necessarily mean that a country is also important from a trade perspective. For example, New Zealand and Australia are relatively small producers, but are important exporting countries and Oceania, together with the EU, are the two key players dominating world dairy exports.

Dairy farming requires a high level of activity-specific investment. As an example, a farmer's livestock is mostly not a natural resource, but is an asset resulting from targeted investment. However, breeding cattle requires the use of natural resources, such as land and water. The livestock asset has an earning capacity resulting from multiple outputs (milk, beef), which in the end also determines its economic value. However, from a sustainability perspective some negative effects have to be acknowledged: dairy cows create negative effects on the quality of the air (e.g. through greenhouse gas emission) and may indirectly put pressure on biodiversity. Other specific investments are in dairy

---

[1]  Organisation for Economic Co-operation and Development. Its membership comprises the EU countries, Australia, Canada, Israel, Japan, Korea, New Zealand, Norway, Turkey and the USA.

**Table 4.1  World dairy production: present and future (million tonnes).**

| | 2009 | 2010 | 2011 | 2012 | 2013 | 2014 | 2015 | Growth %* |
|---|---|---|---|---|---|---|---|---|
| World | 683.2 | 694.1 | 711.1 | 727.6 | 743.4 | 758.9 | 774.7 | 2.1 |
| OECD | 309.2 | 308.9 | 312.8 | 315.3 | 317.7 | 320.0 | 322.9 | 0.7 |
| Non-OECD | 374.0 | 385.2 | 398.4 | 412.3 | 425.7 | 438.9 | 451.8 | 3.2 |
| Developed countries | 358.0 | 357.9 | 363.1 | 367.5 | 371.4 | 375.4 | 379.9 | 1.0 |
| Developing countries | 325.2 | 336.2 | 348.0 | 360.1 | 371.9 | 383.5 | 394.8 | 3.3 |
| Least developed countries | 24.3 | 25.3 | 25.9 | 26.8 | 27.7 | 28.6 | 29.4 | 3.2 |
| European Union-27 | 147.0 | 146.5 | 147.6 | 148.3 | 148.5 | 148.8 | 149.8 | 0.3 |
| Canada | 8.5 | 8.5 | 8.7 | 8.7 | 8.8 | 8.9 | 8.9 | 0.7 |
| United States | 85.8 | 85.4 | 87.0 | 87.7 | 88.7 | 89.6 | 90.6 | 0.9 |
| Australia | 9.7 | 9.2 | 9.3 | 9.6 | 9.8 | 10.0 | 10.1 | 0.7 |
| New Zealand | 16.7 | 17.1 | 17.6 | 17.9 | 18.1 | 18.4 | 18.6 | 1.8 |
| Mexico | 10.8 | 10.9 | 11.0 | 11.2 | 11.3 | 11.4 | 11.6 | 1.1 |
| Japan | 8.0 | 8.0 | 8.0 | 8.0 | 8.0 | 8.0 | 8.0 | 0.1 |
| Turkey | 12.3 | 12.9 | 13.2 | 13.5 | 13.9 | 14.4 | 14.7 | 3.0 |
| Russia | 32.4 | 33.1 | 33.9 | 35.0 | 35.6 | 36.1 | 36.5 | 2.0 |
| Ukraine | 11.1 | 11.4 | 11.5 | 11.6 | 11.7 | 11.9 | 12.0 | 1.4 |
| Egypt | 4.7 | 4.9 | 5.0 | 5.2 | 5.4 | 5.6 | 5.7 | 3.3 |
| Sub-Saharan Africa | 21.2 | 21.9 | 22.5 | 23.1 | 23.9 | 24.6 | 25.2 | 3.0 |
| Argentina | 9.9 | 10.4 | 10.8 | 11.2 | 11.5 | 11.9 | 12.2 | 3.5 |
| Brazil | 28.6 | 29.5 | 30.3 | 31.1 | 31.9 | 32.6 | 33.3 | 2.6 |
| China | 33.3 | 36.7 | 39.6 | 42.0 | 44.1 | 46.2 | 48.4 | 6.4 |
| India | 108.8 | 112.1 | 115.9 | 120.0 | 124.0 | 127.8 | 131.6 | 3.2 |
| Iran | 7.8 | 7.9 | 8.1 | 8.2 | 8.4 | 8.6 | 8.7 | 1.8 |
| Pakistan | 37.3 | 38.6 | 40.1 | 41.6 | 43.2 | 44.7 | 46.3 | 3.7 |

*Average growth 2009–2015.
Source: OECD–FAO, *Agricultural outlook 2010*.

facilities (barns, milking parlours, pasture land, etc.). Together these investments give dairy farms a relatively high fixed cost structure. Moreover, as these costs often have a sunk character, they limit the dairy sector's ability to quickly adjust to changing market conditions.

Moreover, the current era can be qualified as a fast-changing world. We see changing preferences of consumers, an increasing role for environmental and animal welfare groups in setting policy, increasing demand for dairy products as result of population growth and the emergence of economies such as the BRIC countries (Brazil, Russia, India and China). All these developments challenge the resilience and sustainability of the dairy sector.

### 4.1.2 Purpose and focus

The purpose of this chapter is to investigate the resilience and sustainability of the dairy sector in a changing world. For that reason a framework is developed which brings together the most important drivers. Concepts such as resilience and sustainably can be broadly interpreted and have various aspects (environmental, ecological, agronomic, etc.). However, in this analysis the main focus will be economic. Since dairying is usually a commercial activity, the economic aspect is worth studying on its own. The interaction of economic sustainability and other aspects of sustainability will be also dealt with. To analyse the level of resilience and sustainability, different criteria will be used. Two key concepts associated with economic sustainability are profitability and competitiveness. These concepts will be further explored and their sustainability implications will be assessed. Although discussion of concepts plays an important role, empirical applications will also receive attention.

Whereas the concepts and ideas discussed can be applied to every dairy system, they will mainly be illustrated using examples and cases from the European Union. The EU is not only the world's no. 1 dairy region, with dairy being an important activity in nearly all EU member states, but it also shows great heterogeneity. For example, in Poland smallholder dairying is a prominent activity. While in general, in the new member states, dairying is dominated by family farms, several large-scale operations are run as corporate businesses (e.g. in the Czech Republic). In Ireland dairying is a pasture-based activity, whereas in Emillia Romagna (Italy) dairy cows usually have no access to outdoor grazing. In some regions dairy production is highly intensive, with a high cow density per hectare, and problems with manure surpluses (e.g. the Netherlands, Bretagne, Basse Normandie). In disadvantaged (mountainous) areas (e.g. Franche-Comté) cow densities are low and there is a threat that they might further decline below the level necessary for good land, nature and landscape management. The EU dairy sector is heterogeneous not only in terms of production systems, but also in terms of demand for dairy products. The EU faces a growing internal demand, in particular for cheese and fresh dairy products. Dairy products may be differentiated due to policy regulations (e.g. domestic origin protected products) or via commercial brands and voluntary certification schemes. More-over, particularly in the wealthier parts of the EU, consumers are keen for dairying to satisfy increasingly strict criteria with respect to sustainable land use, regulated environmental impacts (e.g. Nitrate Directive) and animal welfare.

In terms of usage, there is no single meaning or definition of the concept of sustainability. The concept carries a variety of meanings, exemplified by the very different interpretations given to it by economists and ecologists. To structure the analysis of sustainability of the dairy sector, it is useful to provide a simple classification of the concept. In our view sustainability encompasses three dimensions: economic, social and environ-mental. These dimensions are also designated as profit, people and planet ('triple P'). As indicated earlier, in this research the focus will be on economic sustainability, but it is acknowledged that real sustainability requires a simultaneous realisation of economic, social and environmental-ecological criteria.

### 4.1.3  Sustainability and dairy

The dairy sector, and agriculture in general, faces three key chal-lenges: the need to produce more in order to feed a growing world population, to produce something different (adjust to consumer demands for food and new services) and, last but not least, to produce better (in respect of the environment, ecology and efficient resource use). The latter challenge is often the first to be associated with sustainability, although sustainability comprises not only respecting the environment (planet), but also includes social (people) and economic (profit) dimensions. As regards the environmental side, the main issues at dairy farm level include the following (Hind, 2010: 3).

- Soil: maintenance of soil structure and productivity, avoidance of erosion, contamination (e.g. residues from plant protection products, heavy metals) and over- and under-fertilisation with organic and artificial fertilisers;

- Air: reducing the level of ammonia and greenhouse gases that come from dairy cattle;

- Water: protect drinking water from diffuse water pollution and contaminants associated with dairy farming, while sustaining access to and availability of fresh water;

- Ecology: protection of habitats from intensive dairy farming activities and ensuring and preserving biodiversity in the farmed environment;

- Landscape preservation: contribute to maintaining an open, pasture-based landscape with typical characteristics.

According to the Millennium Ecosystem Assessment (2005), 10 to 20% of the world's grassland is degraded, which is mainly caused by improper use for livestock grazing. Future growth rates of livestock output will require comparable growth in feed concentrate input. Without taking into account the necessary precautions, further intensification of dairy production can easily result in further land degradation (soil erosion, water pollution). As regards air and climate, the livestock sector is known to be a substantial contributor to ammonia and greenhouse gas (GHG) emissions. According to Gerber and Steinfeld (2010: 6) about 70% of the total ammonia emissions is related to the livestock sector. GHG emissions arise from dairy animals (e.g. enteric fermentation of methane and nitrous oxide emissions from manure) as well as from feed production (chemical fertiliser use, deforestation for pasture, loss of soil organic matter), energy use (e.g. use of fossil fuels in tractors, energy for milking equipment), and the transportation of farm inputs and outputs. Livestock relies directly (drinking and servicing) or (even more important) indirectly (irrigation of feed crops and pastures) on water in order to stay alive. At the same time, water quality can be affected by livestock as a result of the release of nitrogen, phosphorus, pathogens and other substances into waterways. With respect to biodiversity, livestock and wildlife interact in grazing areas. Depending on the stocking density, examples are known where dairy contributes to the improvement of biodiversity, but in more intensive systems, livestock is likely to negatively affect habitats and biodiversity. The standards that are required in order to take care of the environment are likely to differ from place to place, depending on the characteristics of the local environment and the characteristics of the farming system dominating local dairy production.

Although soil, air, water and ecology (nature and wildlife preservation, landscape) are important issues when discussing sustainability, they are not the only issues by far, and often also not the most immediate, that a dairy farmer faces. There are also sustainability issues associated with the dairying production process itself. Farmers are challenged to produce a safe and high-quality product that should be produced on-farm from healthy animals under generally accepted conditions. This imposes increasing demands on the management capabilities of farmers in the areas of:

- animal health;

- milking hygiene;

- animal feed and water;

- animal welfare;

- land or forage area management.

In particular, with respect to this latter list of farm management issues, there is a clear link between sustainability and the economics of dairying. Dealing adequately with many sustainability issues may lead to the imposition of constraints on dairy farming (which then often translates into additional costs of production) and may thereby create a tension or trade-off between the 'planet' dimension and the 'profit' dimension of sustainability. This may also hold true to a certain extent with regard to process standards, but issues concerning the sustainability of the dairying process may also involve win-win possibilities (e.g. avoiding waste).

The International Dairy Federation, jointly with the Food and Agriculture Organization, has developed a *Guide to Good Dairy Farming Practice*, which covers the various sustainability issues we have mentioned (IDF, 2004). For some of the areas identified above there are defined control points that must be managed to achieve defined outcomes. Good agricultural practice (GAP) also implies transparency: dairy farmers should ensure that appropriate records are kept, especially those that enable adequate traceability of the use of agricultural and veterinary chemicals, the purchase and use of animal feed and the unique identification of individual animals. Moreover, records should also be kept of milk storage temperatures (when available) and veterinary or medication treatments of individual animals. People undertaking and supervising the milking operations and management of the dairy farm should be appropriately skilled in animal husbandry, the hygienic milking of animals, the administration of veterinary drugs, the activities undertaken on the dairy farm in relation to food safety and food hygiene, and health and safety practices relating to dairy farm operators (IDF, 2004: 2).

## 4.2   Dairy economics and sustainability

### 4.2.1   Sustainability and resilience of the firm

A common description of sustainability is the ability of a system, a firm or a sector to survive in the long run. The concept of resilience is related to this: in particular, it indicates the ability of a system, firm or sector to maintain its structural and functional capacity after a disturbance or shock (Perrings, 1998: 510). Such a shock might result from changes in physical conditions (e.g. weather shock), market conditions (e.g. declining prices and increased price volatility) or policy changes (e.g. reforms of the EU's dairy policy, including introduction and abolition of milk quotas). Resilience is evidenced by an ability to recover and persist. According to Garmestani et al. (2006: 539) the most resilient industries will be those with functions spread across the range of firm size. The opposite of resilience is vulnerability. The concept of resilience is strongly related to evolutionary theory. According to evolutionary theory, firms (and farms) are able to survive only if they change appropriately over time in response to changes in demand and technology (FitzRoy et al., 1998: 8). Jansen and Osnas (2005: 95) argue that a system is capable of surviving if there is a certain level of redundancy. Redundancy enables a system to maintain its function when a component is lost, and the redundant component takes over that function (Jansen & Osnas, 2005: 95). This means there can be a trade-off between redundancy and efficiency. Farms with a high level of family labour, owned land and personal equity capital have a certain level of redundancy, and are fairly well able to withstand shocks. They can be considered as endowments of the farm household. Business holdings with higher levels of endowments are more able to take risks without ultimately threatening the ability of the business to maintain itself (Phillipson et al., 2004: 229). Other reasons that favour a system being capable of survival, are the following:

- The existence of a certain modularity, i.e. a system has different functional parts or modules that can evolve some-what independently. Applied to dairy farms, certain part of a dairy farm can be outsourced to other farms or businesses in the supply chain (Jansen & Osnas, 2005: 95);

- Diversity in functions of farmers and type of farmers on sector-level adaptation capacity:

  o Adaptive capacity functions at different scales within a system. When the unit of analysis is a social agent, which may vary from an individual to a state, learning, innova-tion and memory are crucial for the resilience of the system;

  o Losing adaptive capacity can be a result of a change in the spatial or institutional environment (see Jansen & Osnas, 2005: 97);

  o The adaptive responses of firms to economic downturn are closely interwoven with the endowments and behav-ioural dynamics of the supporting farm household (Lobley & Potter, 2004; Phillipson et al., 2004: 228);

  o The close relationship between firms and households might on the one hand hinder positive business develop-ment but on the other hand increase the relative viability of production systems (e.g. the family farm);

- External/business environment: small firms have more limited ability to shape their external environment than larger firms. In other words they have to adapt or to adjust to external environmental conditions (Phillipson et al., 2004: 230);

- Networks: The role of personal networks for offering levels of trust is related to the embeddedness argument of Granovetter (1973, 1985). This argument stresses the role of personal relations and structures (or 'networks') of such relations in generating trust and discouraging malfeasance (Granovetter, 1973, 1985), but can be easily extended to link-ages with the supply chain and the way this is organised.

### 4.2.2 *Profitability and the family farm*

A key indicator in measuring the economic sustainability of an activity is profitability. An activity with non-negative (preferably positive) profits is one where the revenues are at least as large as (or larger than) the costs of production. If profits are negative, the revenues cannot cover the costs, which after some time will lead to bankruptcy of the firm and its closure. Positive profits as such reflect that an economic activity adds value, that what is produced is valued more highly by society than the inputs used for its production.

A central assumption in the (micro-)economic theory of producer behaviour is that producers maximise their profits. This is not to say that a producer might not, when asked, mention a set of other objectives which, they argue, also motivate them. This is surely the case but, in economic terms, other objectives can never be pursued while neglecting the profitability issue. Firms compete for resources as well as for clients for their products. This competitive process will keep the drive for profits by individual producers in control. In order to underbid competitors, the margins between revenues and costs will show a tendency to decline until a level is reached which can be labelled as 'normal profits'. These 'normal profits' represents remuneration for the management qualities and skills of the entrepreneur. More generally, economic theory argues, in the long run a market equilibrium will be established where profits tend to zero and the price of (dairy) products will reflect their average cost of production.

An important implication of the economic definition of profitability is that all inputs and outputs, as well as all production factors (labour, land and capital), have to be valued at their 'opportunity' costs. The opportunity cost principle implies that all inputs and outputs, whether they receive a market price or not, will be valued at the market value they have in their next best alternative. For example, a farmer may use a piece of land in his dairy operation. In the economic profit calculation a cost will be charged for the use of this land which is equivalent to the costs associated with renting the piece of land for one year (the lease price), even if leasing does not reflect the real situation a farmer faces (e.g. he might use his own land). The basic justification for this method of calculation is that, when farmers are assumed to go for maximising profits, they should weigh the current use of their resources against any alternative use.

While the reasoning applied above is rational from a pure profit maximisation perspective, when considering this behavioural hypothesis in respect of farmers' behaviour, as can be observed in family farm-type dairy farms, this approach needs further qualification. This may have several reasons, such as farm families having a broader focus than profits only (e.g. following utility maximisation rather than profit maximisation) and/or farmers not following maximising but rather satisficing behaviour. There is no unique definition of what a family farm is, but the following characteristics are typical. The farmer and his family are the main suppliers of the labour input that is needed on the farm. The farm family is usually to a large extent the owner of the means of production (labour, land and equity capital). As a result, the remuneration the farm family receives consists of several parts (i.e. remuneration for labour input, earnings as a capital owner, rents due to the supply of the land, and profit as a compensation for the entrepreneurial and management input the farm head has delivered). Although it is possible to conceptually distinguish all these parts, the farmer will generally receive all this in one amount. This implies that what farmers see as profits or income might be different from the profit calculation that an economist would make.[2] Moreover, as already suggested, farmers and/or their families will most likely follow a household utility optimisation approach, which besides profits also includes other objectives (e.g. handling risk, etc.). This may be even more the case because, for the farm household, consumption and production issues can be intricately interrelated. For example, the farm not only earns them a living, but provides them a home in the countryside which they may greatly enjoy. Also their 'pension scheme' may be linked to their farm (i.e. its asset value). Moreover, their preferred lifestyle may be closely linked to their farm operation. As a result, farmers and their households may extract psychic and/or various forms of in-kind 'income' (utility) from their farm.

---

[2]   To increase the complexity further, a farm family living on the farm can also be stated to receive an income 'in kind' (say, free housing), which is often a big expenditure item for most other households. Moreover, their pension provision is often included in their farm. Thus, aside from income, what happens to the asset value of the land, for example, is also important to properly take into account (wealth effect).

For a family farm dairy farmer, whether his farm is profitable will probably depend on circumstances that economists do not, or only partially, consider. A farmer who from his returns is able to pay all his bills, and still has sufficient money left for a decent living, may evaluate his farm as being profitable. However, placing a value, alongside the paid costs, on the 'unpriced' inputs like owned land and family labour at their imputed economic value might drastically change the picture. The key difference between the economist's perspective and the farmer's view on profits is that the economist looks at the maximum profits possible, whereas the farmer looks at making 'enough' income (see Levins, 1996: 24). The idea of 'enough' is difficult for economists, because its interpretation usually depends on personal (non-objective) circumstances. For family farmers or family business operators in general, the idea of 'maximum profits' is often difficult to grasp, because they see themselves primarily as farmers and not as investors managing a portfolio of resources.[3]

The goal of 'enough' will not only vary from farm to farm, but is also likely to vary for the same farm if circumstances change. There is a kind of a farm life cycle, with young farmers often being heavily indebted, but as they become older their equity increases, and as they come close to retirement they may even decide to disinvest if they have no successor.[4] The focus on 'enough' rather than on maximum profits has two important implications with respect to the resilience and options for sustainability of family farms:

(1)  Since 'enough' will be less than the earnings associated with profit maximisation, the family farm has an advantage as compared to the corporate farm. The latter usually has to follow a profit maximisation approach or an approach close to that, since the inputs and production factors it uses will have to be paid for at market rates (e.g. the use of hired labour

---

[3]  Deviations from the maximisation approach are not specific for agriculture or the dairy sector.

[4]  A distinction has to be made between the farm and the farmer. The life of the farm as a firm hinges on the decisions of the farmer as a firm operator. What a retiring farmer still considers as 'enough', his successor might perceive in a different way. As a result, what for the aged farmer might be a reason to continue his farm for a few additional years might be a reason for a young farmer's son to decide not to become a farmer. As regards disinvestment, see also the remarks below on depreciation.

versus the use of own family labour). While a firm according to the maximising profits approach is making a loss, and thus will have to close its operation if the loss continues, the family farm can accept such a loss, as long as the remaining returns are perceived by the farmer as still being 'enough' or adequate.[5] As a consequence, family farms are expected to be more viable or resilient in very competitive environments.

(2)   What is important from a broader sustainability perspective is that the 'enough' approach leaves open the possibility to concentrate on other goals than solely on profits. Farmers can undertake to satisfy certain standards, even if this involves a cost, and substitute some profits for being able to adhere to a certain farming practice. The space for manoeuvre for corporate-type farms may be much less, since they will be tied more strictly to the profit maximisation approach, which enforces them to minimise all avoidable costs, including costs associated with any standards beyond the generally accepted minimum level that all farmers are expected to adhere to.

Attention has been paid to revenues and costs (profits), in particular the costs associated with owned resources and family labour. In the context of sustainability, it is also important to mention the depreciation costs category. Depreciation is the loss in (economic) value of the capital included in the farm. A sustainable farm needs to be able at least to preserve its capital stock, and by that being able to replace outdated capital goods (machinery, buildings, etc.). From an accounting perspective, depreciation represents a cost but not necessarily an expenditure. However, a farmer who mistakenly sees depreciation as something that can be used for his consumption may disinvest in his farm and may thus damage its sustainability.

### 4.2.3   *Competitiveness*

Sustainability and resilience are concepts that can be assessed at farm level, as was done in the previous section. However,

---

[5]   Negatively formulated, one could say that the family farm mode of production allows for self-exploitation, i.e. farmers accepting for a longer period a remuneration which is below the market 'wage'.

farms always operate in a market environment, which is dynamic and subject to shocks which may have a transitory or a permanent character. Sustainability, seen from an economic perspective, is for that reason also linked to being able to keep and/or improve one's position in the market. This is equivalent to being able to serve consumer needs at a competitive price, or being competitive. As such, competitiveness is closely linked to a sector's or a firm's ability to adjust or restructure itself in response to new and changing environments (Lobley & Potter, 2004; Gardner, 2006).

Surprisingly, the notion of competitiveness – although often used in business and policy-makers' language – has no single definition or clearly established link to economic theory. In itself, competition is a complex economic phenomenon, which alongside the notion of classical price competition includes a multitude of other dimensions. The competitiveness concept has been used in a broad set of contexts and levels of aggregation (country, industry, firm) and is often defined relative to its use.

From a producer's perspective competitiveness may be described as the ability to supply goods and services in the location, form and place sought by buyers, at prices that are as good as or better than those of other potential suppliers, while earning at least the opportunity costs of the return on resources employed. Alternatively, national sector-level competitiveness refers to the ability of a country to produce goods and services that meet the test of foreign or world market competition, while simultaneously maintaining and expanding domestic real income (Kaspersson et al., 2002).

The competitiveness of a firm or sector cannot be separated from the performance of up- and downstream industries. To assess the competitive potential of a complete supply chain, Porter's (1990) diamond framework is helpful. Porter relates competitiveness to the ability to successfully innovate. His approach is thus useful for application to differentiated products. His competitiveness indicator consists of six elements: firm strategy and rivalry, government, demand conditions, related and supported industries, factor conditions and change.

Operational indicators for identifying and measuring the degree of competitiveness include absolute and relative differences in costs of production. Competitiveness is also related to the profitability indicator, discussed in section 4.2.2 above: a profitable dairy operation will be able to maintain and maybe even expand its position as a supplier in the market. Just as with the discussion on profits, on the issue of whether or not to include the opportunity costs of owned resources (land, unpaid labour and other capital invested) to the costs of production, calculations are important. As an indicator of competitiveness, comparing cash costs (excluding the imputed costs associated with owned resources) of milk production can be used to give insight into the dairy sector's or a farm's competitiveness in the short to intermediate run. In the long run, it can be argued that the total economic costs (including costs associated with owned resources) are the appropriate benchmark, since in the long run a sector or a farm is only viable if it succeeds in covering its full economic costs. If not, resources will gradually flow out of the sector and it will have to shrink its production and its market share.

Apart from these indicators (profitability, cash costs and total economic costs of production), insight in the factors explaining these indicators is needed. One way to gain further insight into this is by unpicking the competitiveness indicator in such a way that a number of meaningful indicators – (partial) productivity, intensity and scale – result. For example, the average cost per kg of milk ($AC_j$) can be rewritten as the total costs attributed to dairy ($TC_j$) divided by the total amount of milk ($Y_j$) produced at farm $j$. However, the latter term can be rewritten as the product of the milk yield per dairy cow (*yld/dc*), the number of dairy cows per hectare of forage area (*#dc/ha*) and the number of hectares of forage area available at the farm (*#ha/farm*). See Equation 4.1 (where farm subscripts are suppressed for the sake of convenience).

$$AC = \frac{TC}{Y} = \frac{TC}{\frac{yld}{dc} \cdot \frac{\#dc}{ha} \cdot \frac{\#ha}{farm}} \tag{4.1}$$

The ratios in the denominator have the following interpretation: *yld/dc* represents dairy cow productivity, *#dc/ha* represents cow density per unit of land, and *#ha/farm* is an indicator of the farm size (scale). Together the latter two terms of the denominator can be written as *#dc/farm*, which can itself be interpreted as a herd size indicator (scale). Moreover, combining the first two terms in the denominator, one gets an indicator measuring the amount of milk produced per hectare of forage areas (e.g. $yld/dc.\#dc/ha = y/ha$, with $y = yld/dc.\#dc$), which is an indicator reflecting the intensity of production.

Equation 4.1 can be rewritten as percentage changes:

$$\dot{AC} = \dot{TC} - \dot{Y} = \dot{TC} - \left[ \left( \frac{\dot{yld}}{dc} \right) + \left( \frac{\#\dot{dc}}{ha} \right) + \left( \frac{\#\dot{ha}}{farm} \right) \right] \qquad (4.2)$$

where a dot above a variable indicates its percentage change. As Equation 4.2 shows, the percentage decline in the per unit (or average) costs per kg of milk is the result of the percentage change in its total costs (TC), less the sum of (i) the percentage change in the milk yield per dairy cow, (ii) the percentage change in the number of dairy cows per hectare and (iii) the percentage change in the forage area at the farm. Note that, all else remaining constant (ceteris paribus), increasing the milk yield per cow or the farm size will make the farm more competitive since it then lowers its cost of production per unit of milk.

Whereas Equation 4.2 refers to the change in the absolute costs of production, competitiveness is primarily defined in terms of relative changes in the per unit costs of production (i.e. the change in the costs of production per kg of milk in farm or country A relative to the costs in farm or country B). When this comparison is made between countries that have different currencies, for one country the amount has to be converted into the same currency unit as that of the other country by multiplying the costs by the (appropriately defined) exchange rate. This implies that the relative competitive position of a country's dairy sector not only depends on indicator variables reflecting the performance and structure of the dairy sector, but also the exchange rate, a variable whose value will be mainly determined outside the control of the dairy or agriculture industry (reflecting macro-economic conditions and policies).

In section 4.3, the framework and indicators on economic sustainability and resilience defined above will be applied to the EU dairy sector. The costs associated with achieving minimum standards with respect to soil, traceability and animal welfare will also be dealt with.

## 4.3   Sustainability evaluation of the EU dairy sector

### 4.3.1   Economic sustainability (profit)

Table 4.2 provides an overview of the profitability of dairy farms in the EU (for selected regions[6] and member states with high, average and low gross margins). The data come from the Farm Accountancy Data Network (FADN) database and relate to

**Table 4.2   Profitability of dairy farms in the EU: percentage of specialised dairy farms having positive economic profits,[†] and gross margin over operating costs (€/tonne of raw milk), including its spread.[*]**

|  | 2003 | | 2004 | | 2005 | | 2006 | | 2007 | |
|---|---|---|---|---|---|---|---|---|---|---|
| EU-15 | 33% | 134 | 37% | 130 | 46% | 122 | 44% | 114 | 55% | 148 |
| IT (high) | | 193 | | 187 | | 182 | | 189 | | 199 |
| DE (avg) | | 117 | | 115 | | 105 | | 106 | | 158 |
| FL (low) | | 109 | | 93 | | 70 | | 69 | | 83 |
| EU-10 | | | 49% | 94 | 45% | 108 | 50% | 103 | 56% | 121 |
| PL (high) | | | | 103 | | 117 | | 117 | | 139 |
| CZ (avg) | | | | 81 | | 89 | | 79 | | 78 |
| SK (low) | | | | 49 | | 54 | | 18 | | 83 |
| EU-2 | | | | | | | | | 30% | 173 |

[†]Specialised dairy farms are farms which obtain at least two-thirds of their revenue from dairy operations. Economic profit is defined as the amount remaining when all costs (including remuneration for all production factors) are subtracted from the revenues (including the balance of subsidies such as direct payments).
[*]The spread in gross margin is measured by presenting for each region the member state with a relatively high, average and relatively low value.
IT, Italy; DE, Germany; FL, Finland; PL, Poland; CZ, Czech Republic; SK, Slovakia.
*Source*: FADN data.

---

[6]   EU-15 represents the old member states (situation before 2004 enlargement), EU-10 represents the new member states accessing the EU in 2004, and EU-2 member states (Bulgaria and Romania) entered the EU in 2007.

specialised dairy farms in the EU as prepared by the RICA unit of the EU Commission (see EC, 2010, for more details). Reallocation of costs to the dairy sector is based on output or livestock shares.

As Table 4.2 shows, over the period 2000–2007 a substantial proportion, on average 55% of specialised dairy farms, had positive profits.[7] However, a significant proportion of the specialised dairy farms made a loss, which for some member states and for some years was over 80% (Denmark, Finland, Sweden and Slovenia) (not reported in Table 4.2). Over time the proportion of specialised dairy farms earning a positive profit has been increasing. This is mainly driven by the increase in demand for dairy products relative to supply.[8] Not only has the gross margin increased over time, but the proportion of farms achieving non-negative profits has increased over time. As was argued above, under certain conditions family farms can, at least for some time, have negative profits and still continue their operation. Figure 4.1 shows profitability as approximated by the gross margin (revenue over operating costs). The figure indicates the average value as well as the spread in the gross margin for the EU-15, the EU-10, EU-25 (= EU-15 + EU-10) as well as the EU-27's total average (including Bulgaria and Romania). This indicator can be interpreted as a proxy for profits as evaluated by a cash approach to costs. As Figure 4.1 shows, the gross margin for the EU-15 declined over the period 2003 to 2006, after which it increased in 2007. For the EU-10 the gross margin increased from 2004 to 2005, then it declined in 2006 and strongly increased in 2006. The level of the gross margin in the EU-10 is lower than that realised in the EU-15. The gross margins show a particular variation over member states, with the spread in the EU-15 being much larger than that in the EU-10.

An alternative indicator, which avoids the estimation of imputed remunerations for quasi-fixed production factors and comes closer to actual behaviour, is the critical milk price

---

[7]  Since in general the relatively larger farms have positive economic profits, the proportion of milk produced from dairy farms achieving a positive profit is in general higher than would be expected according to the number of farms.

[8]  With the 2003 policy reform the price support part of the EU's dairy policy was transformed into a safety net provision, which now only protects farmers against strong downside price risks (see further details on the EU's dairy policy in section 4.4).

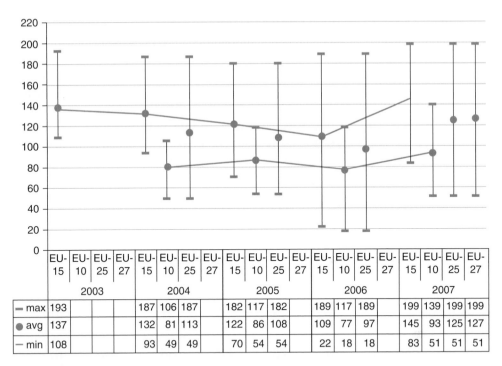

| | EU-15 | EU-10 | EU-25 | EU-27 | EU-15 | EU-10 | EU-25 | EU-27 | EU-15 | EU-10 | EU-25 | EU-27 | EU-15 | EU-10 | EU-25 | EU-27 | EU-15 | EU-10 | EU-25 | EU-27 |
|---|---|---|---|---|---|---|---|---|---|---|---|---|---|---|---|---|---|---|---|---|
| | | 2003 | | | | 2004 | | | | 2005 | | | | 2006 | | | | 2007 | | |
| — max | 193 | | | | 187 | 106 | 187 | | 182 | 117 | 182 | | 189 | 117 | 189 | | 199 | 139 | 199 | 199 |
| ● avg | 137 | | | | 132 | 81 | 113 | | 122 | 86 | 108 | | 109 | 77 | 97 | | 145 | 93 | 125 | 127 |
| — min | 108 | | | | 93 | 49 | 49 | | 70 | 54 | 54 | | 22 | 18 | 18 | | 83 | 51 | 51 | 51 |

**Figure 4.1 Profitability as comprised by the gross margin evolution over the period 2003–2007 (€/tonne of milk).** *Source*: based on FADN data (see EC, 2010).

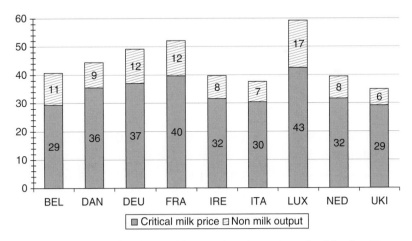

**Figure 4.2 Total critical costs milk as the sum of critical milk price and private consumption in € per 100 kg of milk (average 2006 and 2007) for nine selected EU member states.** *Source*: Jongeneel et al. 2010.

(Figure 4.2). The critical milk price is equal to the milk price a farmer needs to get all his costs (including depreciation) paid and have an adequate living. The latter issue is taken into account by observing the amount of money farmers actually extract from their farm operation for consumption purposes. For the Netherlands in recent years the money extracted for private consumption amounted about €45,000 per farm household, equivalent to about €7.50 per 100 kg of milk (Jongeneel et al., 2010: 17). Note that in countries such as Luxembourg and France the critical milk price is (much) higher than for dairy farmers in the Netherlands (Figure 4.2). The lower the critical milk price the

**Table 4.3  Dairy farms in the Netherlands classified in line with the level of the critical milk price (€/100 kg) in period 2002–2007.**

| Critical milk price range | <25 | 25–30 | 30–35 | 35–40 | >40 | Total |
|---|---|---|---|---|---|---|
| *Characteristics* | | | | | | |
| Number of farms (%) | 10 | 19 | 35 | 21 | 15 | 100 |
| Number of dairy cows per farm | 81 | 78 | 68 | 66 | 44 | 67 |
| Investments/100 kg milk per year | 18.7 | 14.6 | 15.0 | 13.0 | 13.4 | 14.8 |
| *Financial results per 100 kg milk* | | | | | | |
| Returns (incl. subsidies) | 41.30 | 43.10 | 42.00 | 42.50 | 42.70 | 42.30 |
| of which Milk (A) | 32.90 | 33.40 | 33.60 | 33.40 | 33.50 | 33.40 |
| Paid costs and depreciation | 25.00 | 29.50 | 32.20 | 35.90 | 38.30 | 32.10 |
| of which Paid interest | 1.60 | 3.50 | 4.20 | 5.20 | 4.80 | 4.00 |
| Farm income | 16.30 | 13.50 | 9.80 | 6.60 | 4.40 | 10.30 |
| Private consumption | 6.30 | 6.10 | 6.80 | 7.60 | 14.50 | 7.50 |
| Depreciation | 3.50 | 2.70 | 3.20 | 3.10 | 4.30 | 3.20 |
| Redemption | 2.60 | 4.50 | 5.00 | 6.30 | 5.80 | 5.00 |
| Net cash flow (B) | 10.90 | 5.70 | 1.10 | −4.20 | −11.60 | 1.00 |
| Critical milk price (A−B) | 22.00 | 27.70 | 32.50 | 37.60 | 45.10 | 32.40 |

*Source*: Jongeneel et al. (2010).

more competitive dairy farms are. In general it is the case that the calculated critical milk price is below the per unit cost of milk production, accounting for a remuneration of all production factors according to their opportunity costs.

Using underlying data for the Netherlands, Table 4.3 shows that the level of the critical milk price is very different on different farms. Some 10% of Dutch dairy farms have a critical price lower than 25c per kg, while about 15% of dairy farms have a critical milk price higher than 40c per kg. A relatively large group of dairy farms has a critical price around the average of 32.5c. Note that larger farms have on average a 10–20% lower critical milk price than smaller farms (scale economies). Besides the main reason of a lower level of per unit production costs on large farms, another reason is that they 'need' (or accept) less money to cover their accepted consumption level (Jongeneel et al., 2010).

Figure 4.3 provides information about the competitiveness of the EU's dairy sector with respect to a number of key production areas, based on the total costs of production. The information is derived from the IFCN worldwide farm comparison network, in which 134 typical dairy farms from 44 countries were analysed. As the summarising overview shows (see Figure 4.3), Africa is the region with the lowest costs of production (about €14/100 kg), whereas Western Europe (€42.30/100 kg) has the highest costs.

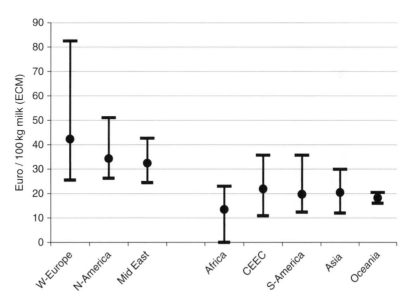

**Figure 4.3   Costs of milk production in dairy inside and outside the EU (euro/100 kg milk; 2007). CEEC, Central and Eastern Europe.** *Source*: Based on IDF (2010).

**Table 4.4  Cost of production and partial productivity, intensity and scale indicators for the EU-15, EU-10 and EU-2 over the period 2000–2007.**

|  |  | 2000 | 2001 | 2002 | 2003 | 2004 | 2005 | 2006 | 2007 | Change* |
|---|---|---|---|---|---|---|---|---|---|---|
| Total operating costs (€/100 kg) | EU-15 | 171 | 181 | 176 | 177 | 179 | 179 | 182 | 201 | 17% |
|  | EU-10 |  |  |  |  | 125 | 134 | 142 | 162 | 30% |
|  | EU-2 |  |  |  |  |  |  |  | 143 | — |
| Milk yield per cow (kg/cow) | EU-15 | 6364 | 6407 | 6531 | 6627 | 6714 | 6819 | 6980 | 7018 | 10% |
|  | EU-10 |  |  |  |  | 5192 | 5367 | 5530 | 5567 | 7% |
|  | EU-2 |  |  |  |  |  |  |  | 3889 | — |
| Dairy cow per hectare | EU-15 | 1.10 | 1.10 | 1.11 | 1.12 | 1.10 | 1.09 | 1.10 | 1.10 | 0% |
|  | EU-10 |  |  |  |  | 0.98 | 0.86 | 0.80 | 0.82 | −17% |
|  | EU-2 |  |  |  |  |  |  |  | 1.46 | — |
| Hectare per farm | EU-15 | 36.68 | 37.61 | 40.03 | 40.99 | 42.91 | 43.64 | 45.97 | 45.76 | 25% |
|  | EU-10 |  |  |  |  | 17.77 | 19.26 | 22.07 | 22.47 | 26% |
|  | EU-2 |  |  |  |  |  |  |  | 3.51 | — |

* % change is 2007/2000 for EU-15 and 2007/2004 for EU-10.

More important for international trade, however, is Oceania which is a low-cost producer (€16.1/100 kg) and produces a large surplus of milk (exceeding the requirements for self-sufficiency by more than 100%) which is exported to the rest of the world, notably Asia.

In order to gain further insight into dynamic changes over time, Table 4.4 shows total costs per unit of milk and the level of milk yield per dairy cow, the number of dairy cows per hectare of forage area and the hectares of forage area per (specialised) dairy farm for the EU-15, EU-10 and EU-2, as well as the percentage change in these variables over the period 2000–2007. Recalling Equation 4.2, the change in competitiveness as measured by the change in total operating costs (cash costs per unit of milk) can be related to different factors. As Table 4.4 shows, the EU-10 is more competitive than the EU-15 (has lower operating costs per 100 kg of milk). However, the EU-10 and EU-15 regions are converging. The EU-15 has a relatively high dairy cow productivity (milk yield per dairy cow) and the intensity of production (dairy cows per hectare) is relatively high compared with the EU-10 (where the indicator declines over time). With respect to farm scale, the EU-15 compares favourably with the EU-10. The relative increase in competitiveness of the EU-15 as compared to the EU-10 can be explained in terms of the relatively strong increase in dairy cow performance and farm scale increase.

### 4.3.2   Environmental sustainability (planet)

Table 4.5 reflects the calculation of costs at the farm and sector level for selected countries for the situation when all affected farms improve their degree of compliance in such a way that they all achieve full compliance with the prevailing nitrate application legislation. Dairy farming in the US does not rely on a pasture system; therefore the acreage per farm is not mentioned (row 3 in Table 4.5). A slightly different procedure was followed for the US and New Zealand case. For selected EU member states, the number of affected and non-affected farms is assessed through farming intensity and location of farms in Nitrate Vulnerable Zones (NVZs) of Europe. Farms with lower densities than 2 livestock units (LU) per hectare ('extensive farms') are assumed to face no significant cost increases. For the full-compliance situation, the percentage increase in total sector costs varies between 0.10% and 4.07%. The absolute cost

**Table 4.5 Dairy and environmental sustainability: estimated costs of compliance associated with satisfying (local) nitrate legislation in five selected EU member states, the US and New Zealand.**

| | France | Germany | Italy | Netherlands | United Kingdom | United States | New Zealand |
|---|---|---|---|---|---|---|---|
| Initial level of compliance (%) | 40 | 60 | 20 | 75 | 95 | 44 | 85 |
| Targeted level of compliance (%) | 100 | 100 | 100 | 100 | 100 | 100 | 100 |
| Average specialised dairy farm size (ha) | 124 | 39 | 28 | 43 | 75 | | 123 |
| Nitrates Directive costs per farm incl. record-keeping costs (€000) | 6.3 | 2.1 | 11.0 | 4.7 | 8.4 | 8.7[†] | 5.9[†] |
| Percentage of specialised affected dairy farms in the sector (%) | 5.0 | 23.0 | 85.0 | 79.0 | 52.0 | 2.0 | 15.0 |
| Percentage increase farm costs (%) | 3.18 | 0.75 | 4.78 | 0.67 | 0.20 | 1.30–4.20 | 3.20 |
| Percentage increase sector cost (%) | 0.17 | 0.17 | 4.07 | 0.53 | 0.10 | 0.02 | 0.48 |

[†] For the US and New Zealand costs refer to those associated with locally applicable measures regulating nitrate use.
Source: Jongeneel et al. (2012).

increases for this group ranges from €400 to €8,800 per farm. It should be noted that many of the less intensive specialised dairy farms as well as the non-specialised dairy farms are assumed to have zero compliance costs. For Europe, the highest percentage cost increase was estimated for Italy, based on a case study carried out for dairy farms of different size and structure exercising various manure management systems (see de Roest et al., 2007).

### 4.3.3  Social sustainability (people)

Sustainability as a wide concept also comprises food safety, transparency and traceability in dairy production. An important part of traceability is good record-keeping of processes and ensuring unique and up-to-date animal identification at the dairy farm. The EU Directives on Identification and Registration (I&R) of animals (92/102/EEG, and Regulations 911/2004, 1760/2000 and 21/2004) aim to ensure this, and represent an important burden, according to farmers (DG-AGRI, 2007). Identification and registration of farm animals is characterised by a significant degree of non-compliance, with 30% non-compliance being not uncommon. The estimated costs for identification and registration compliance at farm level for five selected member states are presented in Table 4.6. It is assumed that there is an eartag loss rate of 15%, requiring proper and timely eartag replacement. The costs due to requirements of the identification and registration standards consist of the eartag costs and the labour (per animal) required for registration.

As can be seen from Table 4.6, the cost per farm associated with identification and registration of animals varies from about €190 to €830. The bottom line of Table 4.6 provides an estimate of the percentage costs of production increase at sector level where it is assumed that dairy farmers will only face additional costs as far as they are not yet compliant with the legislation. As can be seen, this percentage is in general less than 0.1% (Italy being an exception), which implies that the impact on the EU dairy sector's competitiveness due to this requirement will be rather limited.

## 4.4  Agricultural policy

Agriculture, including dairying, is a commercial activity which takes place in a market environment. This market context provides farmers with signals about consumer demands and their

**Table 4.6   Social sustainability (food safety and traceability): estimated costs associated with full compliance with unique animal identification standards in five selected EU member states.**

| Unit | France | Germany | Italy | Nether-lands | United Kingdom |
|---|---|---|---|---|---|
| Initial level of compliance (%) | 90 | 65 | 70 | 90 | 70 |
| Targeted level of compliance (%) | 100 | 100 | 100 | 100 | 100 |
| Average number of animals per specialised dairy farm | 76.5 | 77.7 | 81.8 | 114.2 | 158.6 |
| Estimated number of lost eartags | 11.5 | 11.7 | 12.3 | 17.1 | 23.8 |
| Labour costs per animal (€) | 1.75 | 1.75 | 12.00 | 1.75 | ** |
| Costs tags per animal (€) | 1.80 | 2.92 | 3.00 | 2.75 | 4.20 |
| Total I&R costs per farm (€) | 193.0* | 246.0 | 831.8 | 348.3 | 451.4 |
| Percentage increase sector cost (%) | 0.02 | 0.08 | 0.14 | 0.02 | 0.06 |

* Fixed costs per farm in France are assessed at the level of €9.
** For the UK no specific labour costs were distinguished. They are included in the costs per animal.
*Source*: Jongeneel et al. (2012).

changing preferences, the availability of inputs (e.g. compound feed) and prices as indicators of relative scarcities. However, while no farmer can ignore these signals, there is no assurance that the outcome of the 'free market' will be a dairy production environment that is sustainable. To the contrary, it is quite possible, and even likely, that market signals and the competitive process will place dairy farmers in an economic survival mode, where they may adopt exploitative rather than sustainable behaviour.

However, agricultural policy aims to complement the market system in order to promote outcomes that satisfy a broader range of policy objectives than a narrowly defined efficiency criterion.

The EU dairy policy is part of the Common Agricultural Policy (CAP). Traditionally this policy consisted of a mix of the following policy instruments, which aimed at stabilising the market and supporting farm incomes:

- import tariffs, aimed at supporting the domestic price and protecting the home sector against cheap imports;

- intervention scheme for butter and skimmed milk powder, aimed at stabilising and smoothing the evolution of prices;

- export refunds, aimed at helping disposal of surpluses of dairy products on the world market;

- programmes stimulating domestic demand for specific user groups (e.g. school milk programme).

For many years European import levies and export subsidies were variable in nature: they were adjusted to compensate for the fluctuation in world market prices in such a way as to create a stable price level for dairy products and raw milk within the EU. This stable price level created an environment in which dairy farmers could pursue their investment plans and, as argued by several experts, contributed to the continued modernisation of the sector. The strong focus on price support (and, as compared to the world market, relatively high domestic price level) contributed to further expansion of EU dairy production and to making production more intensive. This contributed to an increasing pressure on the environment, in particular in certain areas where production was concentrated, such as the Netherlands, Denmark, and Bretagne and Basse Normandie regions in France (organic manure surpluses). Whereas the environmental pressure increased with the larger scale, the claims on resources per unit of milk output decreased due to the increasing milk yields per cow (genetic progress).

The EU's classic Common Market Organisation (CMO) for dairy underwent several changes (introduction of milk quota in 1984, change in trade policy due to the Uruguay Round trade agreement, replacement of price support by direct payments, planned quota abandonment in 2015), the most important of which are highlighted below. Even at its inception the EU was close to being self-sufficient in dairy. As production increased over time, the EU became increasingly important as an exporter of dairy products, and together with Oceania is a dominant supplier in the world market. With the increase in exports the budget outlays for export refunds increased, both because of increasing volumes as well as through the depressing effect these exports had on world market price levels. As a result of the latter, the gap between EU and world market prices increased and the per unit export subsidies had to increase. This large increase in budget expenditure on dairy was unsustainable. In response to this financial crisis, in 1984 the EU introduced the milk quota system, which effectively limited the production of raw milk in all member states till the early 2000s, after which for several member states the quota gradually became no longer binding. The milk quota regime is planned to be abolished in 2015 and is being gradually phased out since the Health Check reform of the CAP, which included a so-called soft-landing programme for the dairy sector. This soft landing implies that the milk quotas are gradually increased (by 1% per annum over the period 2008/09 till 2014/15).

The EU's quota or supply management scheme had a favourable impact on the environment. Since supply was curbed while milk yields continued to increase, fewer and fewer dairy cows were needed to produce the quota quantity. As a result the EU's dairy cow herd substantially declined, thereby reducing pressure on the environment (e.g. manure surplus, nitrate and methane emission). Moreover, by imposing zones restricting the tradability of the quota, milk production was retained in regional areas with disadvantaged production circumstances, where otherwise production would have lost its competitive position. In these regions dairy contributed to sustainable land management and the preservation of biodiversity. Moreover, the same territorial restrictions on milk quota movements contributed to limit further concentration of milk production in 'hotspot' areas, which were already coping with significant environmental pressure from animal production.

The milk quota system has been a success in achieving a number of important objectives. Budget outlays for dairy, still one of the most important sectors of EU agriculture, have been substantially decreased over time (with the budget share of dairy declining roughly from 40% in the early 1980s to a little over 4% in the early 2000s). The system contributed to sustaining reasonable prices for farmers, with the income from dairy farming often developing at least as favourably as or even better than incomes elsewhere in agriculture. Moreover the policy contributed, through unintended side effects, to the achievement of regional policy objectives as well as to the improvement of the environmental performance of EU agriculture. This latter 'sustainability bonus' of the milk quota arrangement may be lost when the quota scheme is abandoned in the future, unless the EU's new dairy policy can find other ways to further integrate efficiency, income support and sustainability objectives.

As part of recent reforms (e.g. Agenda 2000, the Fischler Reform of 2003 and the Health Check of 2008) the price support system has changed to a safeguard system, which only protects farmers against extreme downside price risks. Limits have been imposed on the possibilities for intervention and export refunds have been lowered. As a result of the Uruguay Round the EU's variable import levies and export subsidies had to be transformed into a system of fixed rates (tarification). Together these developments implied a greater impact in the EU from price changes occurring in the world market. As a consequence both the level of EU raw milk prices as well as their volatility has increased, in particular since 2005. Dairy farmers were partly compensated for the institutional price declines (reduction in intervention prices to a safeguard level) by direct payments. Initially these payments were still coupled to milk production (e.g. the milk premium), but since 2007 these payments have been included in the Single Farm Payment scheme, and are fully decoupled from production. These payments contribute significantly to the resilience of EU dairy farmers. The decline in milk prices as well as the decoupling of income support from dairy production creates incentives to lower the supply of raw milk and thereby contribute to a less intensive dairy production in the EU. However, these effects seem till now to have been limited, although different from the years before. Because the milk quota has become non-binding in many member states, the pressure to further expand milk production in the EU is, on the whole, not very high.

## 4.5  Conclusion

This chapter has discussed sustainability and resilience in dairying, with a focus on economic sustainability and how this interacts with other sustainability issues, notably environmental sustainability (nitrate) and food safety. As indicators for economic sustainability and resilience, attention was paid to profitability, costs of production and competitiveness. The specific characteristics of the family farm, in particular its resilience relative to comparable corporate farms, were discussed. Concepts were introduced and an empirical application to the EU-27's dairy sector made. The EU dairy sector shows significant heterogeneity and as such provides a good illustrative case, comprising elements that will also be found when applying the methodology to other countries. The picture generated by the set of indicators that is analysed (gross margin, percentage of specialised dairy farms earning a positive net economic profit, operating costs of production, and the critical milk price) is that dairy production is only partially sustainable from a profit maximisation perspective. Whereas a significant proportion of farms achieves a positive profit, there also is a significant proportion earning a negative economic profit. As the spreads in gross margins and the discussion of the critical milk price showed, there is a strong variation across member states and even more so across farms. With respect to dairy farmers, large farms in general perform better than small farms (emphasising the importance of economies of scale). Alongside structural characteristics, such as dairy cow productivity, intensity of production and farm scale, the policy environment was argued to be of importance.

As for the EU, its dairy policy, as part of the CAP, protects the EU dairy sector from foreign competitors. Moreover, the price support policy, which was a keystone of the EU's dairy policy until 2005, has contributed to keeping dairy farms in production which would otherwise have found it difficult to survive. However, this past policy hampered the EU sector's external competitiveness. The recent policy turn from price support to income support via direct payments has improved the EU's external competitiveness, but at the same time has made the sector more dependent on public transfer payments (which are themselves not very sustainable).

The EU dairy sector as such faces several challenges, having at the same time to improve its economic as well as its environmental and social sustainability. As this research shows, the 'triple P' dimensions of sustainability are all interconnected. In particular, since environmental sustainability is likely to increase the costs of production, economic profitability is an important side requirement in order to improve the EU dairy sector's general sustainability. Policies can play an important role in improving the dairy sector's profitability. As compared to the past, income support and environmental sustainability issues need to be better integrated, for example by introducing a better targeting of payments made to farmers (following a green growth-oriented strategy).

# References

de Roest, K., Montanari, C., Corradini, E. (2007) *Il costo della Direttiva Nitrati. BIT SpA – Cassa Padana. Leno (Brescia).* Emillio Romagna: Centro Ricerche Produzioni Animali.

DG-AGRI (2007) *Report from the Commission to the Council on the Application of the System of Cross Compliance.* COM (2007) 147, 29.03.2007, 2007b). Brussels: European Commission.

European Commission (2010) *EU dairy farms report 2010 based on FADN data.* Brussels. European Commission.

FitzRoy, F.R., Acs, Z.J., Gerlowski, D.A. (1998) *Management and Economics of Organization.* London: Prentice Hall Europe.

Gardner, B.L. (2006) Agricultural support policies, productivity and competitiveness. Paper presented at the USDA and AIEA2 International Meeting 'Competitiveness in Agriculture and the Food Industry: US and EU Perspectives'. 15–16 June 2006, Bologna, Italy.

Garmestani, A.S., Allen, C.R., Mittelstaedt, J.D., Stow, C.A., Ward, W.A. (2006) Farm size diversity, functional richness, and resilience. *Environment and Development Economics,* 11: 533–551.

Gerber, P.J., Steinfeld, H. (2010) Global environmental consequences of the livestock sector's growth. *Bulletin of the International Dairy Federation,* no. 443: 4–12.

Granovetter, M.S. (1973) The strength of weak ties. *American Journal of Sociology,* 78(6): 1360–1380.

Granovetter, M.S. (1985) Economic action and social structure: the problem of embeddedness. *American Journal of Sociology,* 91(3): 481–510.

Hind, T. (2010) Overview of the main environmental issues at farm level and the work that has already been done in the guide to good dairy farming practice. *Bulletin of the International Dairy Federation,* no. 443: 3.

International Dairy Federation (2004) *Guide to good dairy farming practice.* Brussels: IDF.

International Dairy Federation (2010) The world dairy situation 2010. *Bulletin of the International Dairy Federation,* no. 446. Brussels: IDF.

Jansen, M.A., Osnas, E.E. (2005) Adaptive capacity of social-ecological systems: lessons from immune systems. *Ecohealth,* 2(2): 93–101.

Jongeneel, R., van Berkum, S., de Bont, C. van Bruchem, C., Helming, J., Jager, J. (2010) *European dairy policy in the years to come: quota abolition and competitiveness.* Report 2010-017. The Hague: LEI-WUR.

Jongeneel, R., Bezlepkina, I., Brouwer, F., Dillen, K., Meister, A., Winsten, J., et al. (2012) Dairy. In F. Brouwer, G. Fox, R. Jongeneel (Eds.) *The economics of regulation in agriculture: compliance with public and private standards.* Wallingford: CABI.

Kaspersson, E., Rabinowicz, E., Schwaag Serger, S. (2002). *EU milk policy after enlargement: competitiveness and politics in four candidate countries.* Lund: Swedish Institute for Food and Agricultural Economics.

Levins, D. (1996) *Monitoring sustainable agriculture with conventional financial data*. White Bear Lake: Land Stewardship Project & University of Minnesota.

Lobley, M., Potter, C. (2004) Agricultural change and restructuring: recent evidence from a survey of agricultural households in England. *Journal of Rural Studies*, 20: 499–510.

Millennium Ecosystem Assessment (2005) *Ecosystems and human well-being: current state and trends*. www.millenniumassessment.org/documents/document.766.aspx.pdf (accessed 27 August 2012).

Perrings, C. (1998) Resilience in the dynamics of economy-environment system. *Environment and Resource Economics*, 11(3–4): 503–520.

Phillipson, P., Bennett, K., Lowe, P., Raley, M. (2004) Adaptive responses and asset strategies: the experience of rural micro-firms and foot and mouth disease. *Journal of Rural Studies*, 20: 227–243.

Porter, M.E. (1990) *The competitive advantage of nations*. New York: Free Press.

# 5

# Dairy processing

Arjan J. van Asselt[1] and Michael G. Weeks[2]
[1] NIZO Food Research BV, Ede, The Netherlands
[2] Dairy Innovation Australia Ltd, Werribee, Australia

**Abstract:** The key unit operations in the dairy industry and their impact on the environment are discussed. Common practice, new technologies and computer tools are presented on how to reduce the use of energy and water and the emission of waste products.

**Keywords:** cleaning, drying, energy, evaporation, fouling, heating, membrane separation, water

## 5.1   Introduction

Milk processing in general can be quite energy and water consuming depending on the end-product and chosen process. During its life cycle milk is heated and cooled several times. At the farm, milk is cooled to a temperature of 4°C. After the milk is received at the factory the milk is heated and cooled more than once (depending on the type of end-product into which milk is transformed, e.g. cheese, powder, yogurt, etc.). After processing the milk will leave the factory at temperatures between 4°C (e.g. fresh products) and 20°C (e.g. powders, ambient stable

*Sustainable Dairy Production*, First Edition. Edited by Peter de Jong.
© 2013 John Wiley & Sons, Ltd. Published 2013 by John Wiley & Sons, Ltd.

milk). The Food and Agriculture Organization has published a life cycle assessment of dairy products (FAO, 2010) and estimated the greenhouse gas (GHG) emissions as 0.155 kg $CO_2$-eq per kg milk from farm gate to storage at the retail outlet; of this, 0.086 kg $CO_2$-eq per kg milk (or 86 kg per 1,000 kg of milk) was derived from milk processing at the factory. These are average values for Europe and vary by region and by product. For example, fermented milk has a quite high emission rate of 0.304 kg $CO_2$-eq per kg milk whereas cheese and whey have a lower rate of 0.126 kg $CO_2$-eq per kg milk at the farm gate. When calculated on a finished product basis, the emission rate for cheese is higher, at approximately 1.260 kg $CO_2$-eq per kg cheese (based on 10 kg milk for 1 kg Gouda cheese). The significant point is that processing contributes a rather small part to the total amount of GHG. The International Dairy Federation calculated in 2009 that 1.2 kg $CO_2$-eq was emitted for the production of 1 kg of fresh milk and 8.8 kg $CO_2$-eq for the production of 1 kg of cheese (IDF, 2009). These figures are based on the whole chain (from farm to retail). From farm gate to retail, processing is responsible for approximately 13–14% of the GHG emission of milk (i.e. 0.17 kg $CO_2$-eq per kg milk). For the USA the equivalent whole-of-chain emission was calculated at 2.05 kg $CO_2$-eq per kg milk, of which 5.7% is caused by processing (i.e. 0.117 kg $CO_2$-eq per kg milk) (US Dairy, 2010). From these reports it can be concluded that the GHG emissions for processing are of the same order of magnitude and that the contribution of processing emissions compared with farm-level emissions is much lower.

With regard to water use per kg of milk, it is more difficult to obtain a reliable view. Data are often not reliable or available in order to make a correct assessment. In addition several types of water use can be distinguished, such as consumptive or degradative (IDF, 2009) as well as the concept of water stress index (Ridoutt, 2010). However, an average amount of direct water use in milk processing is around 1.5 kg of water per kg milk. This water is mostly used for cleaning, heating and cooling. Regional variations are observed from 1.18 L per kg in the UK (Dairy UK, 2009), 1.29 L per kg in Australia (Dairy Australia, 2009), 1.5 L per kg in Denmark (Flapper, 2009) and 2.2 L per kg in New Zealand (Fonterra, 2011). When the whole life cycle is included (from cow to fork), the amount of water is almost 1,000 L per kg milk (IDF, 2009).

From these data it can be concluded that a single specific figure cannot be given for dairy processing and that data will vary by region and by product. However, in order to provide some focus, we can consider processing in terms of the main unit operations that are generally applicable to the whole dairy industry. Based on these data, the industry is able to determine the points of interest (for energy and water reduction) with regard to specific activities. The key unit operations within the dairy industry are described and discussed below.

## 5.2    Key unit operations and their water and energy use

### 5.2.1    Milk pre-treatment

In order to quantify and compare the various unit operations, $CO_2$ equivalents are calculated based on energy use. The conversion factors used originate from the Carbon Trust (Carbon Trust, 2011). In the calculations, electrical energy (kWh) is directly converted into $CO_2$-eq based on the factor for grid electricity. This is the maximum (worst case) value; values might differ if, for example, combined heat and power values are used. For thermal values, natural gas is used as a basis for $CO_2$-eqs.

After the milk is received at the factory the following unit operations take place.

- *Storage*. Milk is stored at temperatures between 4 and 7°C in stainless steel storage tanks. If applicable, tanks have no cooling mechanism. If applicable (i.e. in subtropical and tropical regions), tanks are equipped with insulation in order to prevent warming up of the milk during storage.

- *Standardisation*. Raw milk is standardised for fat and (if applicable) for protein content. Milk is therefore split into cream and skim milk using a separator (centrifuge). Depending on the desired fat (and protein) content the two streams are mixed again in defined proportions. Energy use (electrical) is around 45 kW at a capacity of 40,000 kg/h (GEA, 2010[1]) per kg of milk processed, which equals 0.61 kg $CO_2$-eq per 1,000 kg milk (Carbon Trust, 2011). Applying a load factor of 85% of the installed capacity, this figure reduces to 0.52 kg $CO_2$-eq. Thus, out of the total amount of 86 kg $CO_2$-eq for processing of 1,000 kg milk (FAO calculation, section 5.1), standardisation contributes 0.6%. Water use is determined at

[1]    www.westfalia-separator.com/products/product-finder/product-finder-detail/product/separator-msi-400-01-772.html#technicalData.

350 kg per 40,000 kg of milk processed. Tetra Pak has not published data on its energy and water use in milk processing; however, it is believed to be of the same order of magnitude. The company has developed a hermetically sealed separator resulting in a claimed reduction of energy use of 35%, through moving from gearbox to belt to direct drive machinery. In addition, advances in cleanability have resulted in reduced product loss at discharge, higher solids desludges and reduced water requirements through much shorter discharge times (0.2s discharge) and longer times between discharge (Rowlands, 2011).

- *Homogenisation*. After standardisation, milk is homogenised to reduce the size of the fat globules or particles to prevent or delay the natural separation of cream from the rest of the milk emulsion. To achieve this, milk is pumped under very high pressure (typically up to 250 bar) through a very narrow valve, resulting in disruption of the milk fat globule and thus size reduction. Energy use (electrical) is around 315 kW per 36,000 kg/h at 250 bar, which equals 4.7 kg $CO_2$-eq per 1,000 kg of processed milk. Homogenisation is used applied for fresh and UHT-type milk products (fresh milk, cream, yogurt, UHT and extended shelf-life (ESL) milk) and not for cheese processing. There is a trend within the organic dairy sector not to use homogenisation at all. From a food safety point of view, homogenisation is not strictly necessary, as the process does not result in inactivation of bacteria. Energy savings can be made by processing only a portion of the skim with the cream, thereby reducing the required throughput and power of the homogeniser. The amount of power drawn increases directly with the flow rate and homogeniser pressure. Some advances have been made that enable more energy-efficient homogenisation: for example, the patented NanoValve® from NiroSoavi[2] enables homogenisation in standard milk applications to take place at a lower homogenising pressure and a high flow rate application.

In the pre-treatment of milk, standardisation is the only process that is carried out for all types of milk. Homogenisation is carried out optionally, depending on the size of the factory and on the type of products being produced. The total impact of pre-treatment processes on the carbon footprint is relatively low.

---

[2] www.nirosoavi.com/high-pressure-homogenisation-technology.asp.

### 5.2.2   *Milk heat treatment*

In order to make milk microbiologically and enzymatically stable, it requires preservation. Alternative methods of preservation such as pulsed electric fields, ohmic heating and high pressure processing are under development; however, these have much higher costs of operation, are less robust and not always available on an industrial scale. Heat is still the main treatment applied. Depending on the type of product (fresh or ambient stable) the following heat treatments are used.

- *Thermisation.* The objective should be to heat treat all incoming milk within 24 hours. If this is not possible, thermisation is often used to maintain the quality of milk before final processing. The milk is pre-heated for 10–15s at temperatures between 63°C and 65°C within 1 day after receipt at the factory. After this treatment the milk is still phosphatase positive (as required by legislation). Heat treatment is carried out with plate or tubular heat exchangers. Heating is carried out indirectly using hot water as a heating medium. Energy use (steam or gas) for heating milk up to 65°C is expected to be similar to the amount of energy needed for pasteurisation (between 71.57 and 48.7 MJ/1,000 kg) depending on the regenerative effect (reuse of heat) varying from 90% to 95% (APV, 2011[3]; IDF, 2005). Although the temperature of thermisation is lower, the effective temperature difference to be heated by using primary energy is similar (between 7°C and 3.5°C).

- *Pasteurisation.* Milk is pasteurised in order to inactivate vegetative microorganisms and to obtain milk that is stable in the cooling chain at temperatures between 4°C and 7°C for 10 to 14 days. Minimum heat treatment within the European Union is normally set at 72°C for 15s. Milk intended for cheese making is pasteurised in order to obtain a certain degree of denaturation of whey proteins. This denaturation is relevant with regard to the cheese yield (the more denatured the whey proteins, the higher the water binding and the higher the amount of cheese produced). Total energy use (steam or gas) was calculated as 71.5 MJ/1,000 kg milk for heat exchangers with 90% regeneration (IDF, 2005). Improved

[3]   www.apv.com/us/products/heatexchangers/Gasketed/gasketedplateframehe_apv.asp.

Table 5.1  Typical temperature settings per heat treatment.

| Product/treatment | Pasteurised 100% indirect (plate & frame) | ESL 100% indirect (Sterideal) | UHT 100% indirect (Sterideal) | UHT 86% indirect (Steritwin) | UHT 54% indirect (VTIS, Steritwin+) |
|---|---|---|---|---|---|
| Capacity (kg/h) | 20,000 | 20,000 | 20,000 | 20,000 | 20,000 |
| Temp in | 10 | 10 | 10 | 10 | 10 |
| Regenerative heating out | 65 | 108 | 121 | 113 | 63 |
| Indirect steam heating out | 72 | 125 | 138 | 130 | 80 |
| Direct steam heating out | 72 | 125 | 138 | 150 | 145 |
| Flash cooling out | 72 | 125 | 138 | 130 | 80 |
| Regenerative cooling out | 17 | 27 | 27 | 27 | 27 |
| Cooling/icewater cooler out | 10 | 10 | 20 | 20 | 20 |
| Steam heating dT | 7 | 17 | 17 | 37 | 82 |
| Cooling dT | −7 | −17 | −7 | −7 | −7 |

heat transfer designs for plate heat exchangers have allowed very high regeneration efficiencies to be obtained. When the heat regeneration rate is increased to 95%, energy use decreases to 48.7 MJ/1,000 kg milk processed. If the regeneration efficiency is lower or there is an imbalance in the flow on the heat-up side versus the cool-down side (e.g. due to separation of cream from the milk), then there is a two-fold effect. More steam is needed to achieve the required high heat temperature and also more cooling (chilled water, electrical energy) is required to achieve the final product temperature.

- *UHT and sterilisation.* Various combinations of time and temperature are possible, including variations in heat transfer. Table 5.1 gives an overview of the various UHT treatments and their temperature settings. Based on these settings, the amount of energy (thermal and electrical) is calculated. The energy use for these treatments is shown in Table 5.2. Clearly, the amount of energy used is directly related to the amount of heat regenerated in the system and to the heating temperature applied. The electrical power of the various treatments is more or less equal and depends mainly on whether homogenisation is required (see also section 5.2.1).

### 5.2.3  *Evaporation*

Within the dairy industry, falling-film evaporators are commonly used for pre-concentration in powder production (e.g. for milk powder or infant formula) or for the production of evaporated milk products (e.g. condensed milk). Evaporators were designed to concentrate milk products with much higher energy efficiency compared to drying. The specific energy use for evaporation is in the range of 200–600 MJ/1,000 kg water evaporation whereas for drying the thermal energy requirement is 4,000–4,500 MJ/1,000 kg water evaporated, depending on the configuration (Verdurmen & de Jong, 2000). Two types of evaporator can be distinguished: TVR (thermal vapour recompression) and MVR (mechanical vapour recompression) evaporators. Usually an MVR evaporator is combined with a TVR evaporator: MVR is used for products with a maximum dry solids content of 30–35% and TVR is then only used for the final concentration up to 50–55% dry solids. However, some production facilities (e.g. that of Unipektin) use MVR to concentrate milk from 12% to 50% dry solids. The energy use of MVR evaporators is much lower than that of the TVR type. Figure 5.1 shows the difference

**Table 5.2   Steam and energy consumption of various heat treaments.**

| Product/treatment | | Pasteurised 100% indirect (plate & frame) | ESL 100% indirect (Sterideal) | UHT 100% indirect (Sterideal) | UHT 86% indirect (Steritwin) | UHT 54% indirect (VTIS, Steritwin+) |
|---|---|---|---|---|---|---|
| **Steam consumption** | | | | | | |
| steam consumption | kg/h | 280 | 680 | 680 | 1480 | 3280 |
| Cooling power consumption | kW | 171 | 416 | 171 | 171 | 171 |
| **Electrical power consumption** | | | | | | |
| fluid transport pressure loss | Pa | 5.0E+05 | 1.0E+06 | 2.0E+06 | 2.5E+06 | 2.0E+06 |
| fluid transport power | W | 5.6E+03 | 1.1E+04 | 2.2E+04 | 2.8E+04 | 2.2E+04 |
| homog pressure | Pa | 1.2E+07 | 1.5E+07 | 2.0E+07 | 2.0E+07 | 2.0E+07 |
| homog power | W | 9.5E+04 | 1.2E+05 | 1.6E+05 | 1.6E+05 | 1.6E+05 |
| Total electrical power consumption | W | 1.2E+05 | 1.6E+05 | 2.2E+05 | 2.2E+05 | 2.2E+05 |

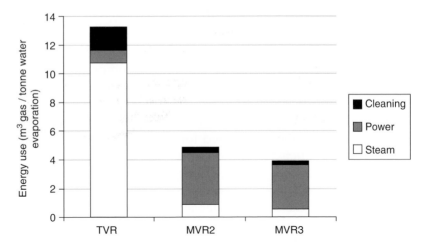

**Figure 5.1    Energy use per type of evaporator: TVR, MVR 2 (second generation) and MVR 3 (third generation) in m³ gas per tonne of water evaporation.**

in energy use between the two types of evaporator. The data show clearly the large difference between the standard TVR evaporator and the most recent generation of MVR evaporators (third generation). The carbon footprint per 1,000 kg of water evaporation for the two types of evaporators is 26.8 kg $CO_2$-eq for a TVR and 7.9 kg $CO_2$-eq for a third-generation MVR (Vissers et al., 2002).

## 5.2.4   Drying

Drying is by far the most energy-consuming operation within the dairy industry. Three different drying technologies can be distinguished: spray drying, drum drying and fluidised bed drying. Spray drying is used in the production of powders with good dissolving properties, such as plain milk powder, coffee whiteners and infant formula. Within a standard milk powder processing plant, 90% of the water is removed in the evaporator and the remaining 9–10% in the spray dryer. However, the energy required per kg water evaporated in the dryer is about 15 times higher than the energy required in the evaporator (combination of TVR and MVR). As indicated in section 5.2.3 the specific energy use for spray drying is 4.4 MJ/kg water evaporation which equals 665 kg $CO_2$-eq per 1,000 kg water evaporation (conversion factors – Carbon Trust, 2011). The main reason for this high energy use in spray drying is the slow heat transfer from hot air to liquid, which makes it necessary to use large amounts of hot air to remove the water from the liquid droplets. In addition the thermal efficiency is modest due to limits on the maximum inlet air temperature and the minimum exhaust air temperature, because of product quality and processability constraints.

A more energy-efficient method of drying is drum drying. Drum drying is used for the more viscous, sensitive and sticky products (e.g. those containing starch) and results in thin flakes (e.g. potato flakes, modified starch, milk powder for chocolate). Compared to spray drying, drum drying is indirect and therefore the heat transfer is more efficient and uses less energy to remove the water. The amount of steam used in drum drying for the removal of water varies between 1.1 and 1.6 kg steam per kg of evaporated water, corresponding to energy efficiencies of about 60%–90%. Based on an energy content of 2.4 MJ/kg steam, this implies a thermal energy use of between 2,640 MJ/1,000 kg and 3,840 MJ/1,000 kg evaporated water and is lower than the 4,400 MJ/1,000 kg for spray drying.

The third type of drying is fluid bed drying. Fluid bed processing involves drying, cooling, agglomeration, granulation and coating of particulate materials. It is ideal for a wide range of both heat-sensitive and non-heat-sensitive products. Uniform processing conditions are achieved by passing a gas (usually air) through the product layer under controlled velocity conditions to create a fluidised state. In fluid bed drying, the fluidisation gas supplies heat, but the gas flow needs not be the only heat source. Heat may be effectively introduced by heating surfaces (panels or tubes) immersed in the fluidised layer (GEA, 2011b[4]). Fluid beds are commonly integrated with spray dryers (internal fluid beds) and/or may be external beds in order to improve the overall thermal efficiency of the drying stage. The thermal energy required decreases from 6.7 MJ/kg for a single-stage skim milk dryer to 4.1 MJ/kg for a compact dryer to 3.7 MJ/kg for a high-temperature multi-stage dryer.

### 5.2.5 *Membrane filtration*

Membrane filtration is generally used for the isolation of valuable ingredients from milk or by-product streams or for the removal of microorganisms. Membrane filtration makes use of a

---

[4]   www.niroinc.com/drying_dairy_food/fluid_bed_technology.asp.

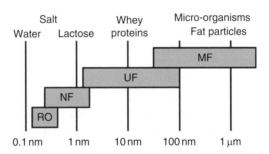

**Figure 5.2    Pore size and type of product isolated with specific membrane applications. RO, reverse osmosis; NF, nanofiltration; UF, ultrafiltration, MF, microfiltration.**

semi-permeable polymeric or ceramic membrane. An overview of pore sizes and types of application is given in Figure 5.2. The main driving force for the separation process is pressure and therefore this determines the energy used. Within this pressure-driven process the following four categories can be distinguished:

1. reverse osmosis (RO, operating pressure 3–4 MPa).
2. nanofiltration (NF, operating pressure 1.5–3 MPa).
3. ultrafiltration (UF, operating pressure 0.3–0.8 MPa).
4. microfiltration (MF, operating pressure 0.05–0.2 MPa).

Normally, centrifugal pumps are used to build up the operating pressure, so that energy use is determined by the pump capacity used and pressure lift required. For a standard RO process with a capacity of $30\,m^3\,h^{-1}$ and a maximum pressure of 4 MPa, a centrifugal feed pump with a motor of max. 10 kW would be sufficient (Bargeman, 2003). In addition, a recirculating pump is needed which has a capacity of approx. 13 kW (based on a one-stage recirculating process). So the power of the two pumps is of the same order of magnitude.

## 5.2.6  *Cleaning*

Cleaning of premises and equipment is an important unit operation that is commonly applied throughout the whole food industry. In the dairy industry, daily cleaning is required for the major part of the equipment. This ensures a constant product quality and efficient heat transfer in heat exchangers, and avoids the possible growth of microorganisms. It is clear that a high frequency of cleaning has a large impact on the availability of the processing equipment and on the use of energy and water. A standard cleaning approach is 'cleaning in place' (CIP), which is achieved by the circulation of hot alkaline and acid solutions. Phase transitions between product and cleaning solution and vice versa are carried out with water. At the end of CIP cleaning, a final rinsing step is carried out using fresh potable water. This water is often reused as the pre-rinse water for the next cleaning. The main energy-consuming part of a CIP process is the heating of the cleaning liquids to the required temperature. Alkaline solutions are normally heated to a temperature of 80°C and stored in a CIP tank. When cleaning is started the cleaning liquid is pumped through the equipment at a temperature of approximately 20°C. This temperature difference implies that the temperature of the cleaning liquid decreases and needs to be raised in order to ensure a minimum cleaning temperature of approx. 75°C, which is necessary from a microbial point of view. For an average heat exchanger (with a capacity of approximately $36\,m^3h^{-1}$) the energy and water use for three different cleaning regimes are shown in Table 5.3.

Table 5.3    Energy use, $CO_2$-eq, water use and emissions for three different cleaning regimes for a heat exchanger in a pasteurisation process (based on NIZO calculations).

| Cleaning regime | Energy use (MJ/ tonne product) | $CO_2$-eq (kg/tonne product) | Water use processing (kg/ tonne product) | Emission (fouling equivalents/ year) |
|---|---|---|---|---|
| Cleaning once every 8 hours | 1.9 | 0.11 | 46.9 | 3848 |
| Cleaning once every 24 hours | 0.6 | 0.04 | 15.6 | 1279 |
| Cleaning once every 24 hours + 1 intermediate clean | 1.1 | 0.06 | 15.6 | 2564 |

These data show that the energy use in cleaning is quite low compared with the energy use in production processes such as pasteurisation. Cleaning has the largest impact on water use. If no cleaning or product change were necessary, little water would be needed.

A different method of cleaning is known as 'cleaning out of place' (COP): COP cleaning involves taking apart sections of equipment and parts of fillers which require disassembly for proper cleaning. Parts removed for cleaning are placed in a circulation tank and cleaned using a heated chemical solution and agitation. Compared to CIP, COP is used only on a very small scale and will not be discussed further in this chapter. A third way of cleaning is manual cleaning (e.g. with high-pressure cleaning equipment), which is normally used on equipment where circulation or dismantling are not possible, for example, cheese presses in a cheese factory, or conveyer belts. It is difficult to give figures for energy and water use since they will be influenced by a series of uncertain factors such as the type of apparatus, personal approach, time, etc. As almost no water or cleaning liquid are circulated or reused in this type of cleaning, it is less water- and energy-efficient than CIP or COP.

Generally, water use in CIP, crate washing and manual cleaning can amount to 50% of the water used on site for a fluid milk plant (Prasad et al., 2005). Fresh water requirements for CIP can be significantly reduced by the adoption of CIP reuse systems including rinse recovery, fit-for-purpose water reuse and effective system design and monitoring.

### 5.2.7 Storage (conditioning, cooling)

Standard energy use for cooling a dairy product and keeping it at a temperature of 4°C is calculated at 0.6 kWh per tonne per day (van den Berg et al., 1983). For example, to keep a pasteurised product in the cooling chain for 14 days at 4°C implies an energy use of 8.4 kWh per tonne, which equals 4.6 kg $CO_2$-eq per tonne.

### 5.2.8 Utilities (heat generation, cold generation)

The main energy-consuming utilities in dairies are refrigeration, steam, hot water and compressed air systems. While electricity is directly utilised for a portion of these processes, the majority of

energy purchased is required for a combination of thermal applications with electricity utilised in secondary energy generation.

Secondary end-use energy is obtained through the combustion
of primary energy sources for all thermal applications in dairies
(i.e. steam and hot water or direct process heat). Refrigeration
and compressed air services are also generated using electricity.
The magnitude of energy lost during the conversion of these
primary energy sources is very much dependent on the design,
configuration and operation of these plant services.

There is potential to improve the conversion efficiencies for
services in all dairies. The following is a review of the types
of systems employed and their relative performances, describ-
ing potential opportunities for optimisation of refrigeration
performance, compressed air, steam and hot water systems. To
provide dairy factories with heat and cold, in most cases each
factory has its own steam boiler and refrigeration system. With
regard to steam boilers, water is heated using gas, oil or coal,
thus producing steam at a certain pressure (normally between 6
and 12 bar) and up to 40 bar for powder plants. In general, well-
tuned boilers have an overall boiler efficiency of approx. 85% for
producing steam at 13.5 bar (abs). For the production of 10,000 kg
of steam (13.5 bar abs, $T = 191°C$), 826 $m^3$ of natural gas is needed
(efficiency 85%). Based on the energy content of natural gas
of 35.3 MJ/$m^3$ gas, this implies 2,915 MJ/1,000 kg steam (2.9
GJ/1,000 kg steam).

For cooling, a standard solution is the application of ice water
(temperature around 1°C). Ice water is produced using a refrig-
erating installation based on evaporation and condensation of a
cooling agent such as $CO_2$ or ammonia ($NH_3$) or artificial refrigerants such as R22. The latter is being phased out from cooling
systems within the European Union. The energy for cooling is
electrical energy, which is used in compression of the vaporised
cooling agent. For fresh milk products the energy needed for
cooling is calculated to be 0.019 MJ/kg (Oldenhof, 2004), based
on a temperature decrease of 5°C after leaving the heating equipment (i.e. pasteuriser).

Refrigeration plants are a significant consumer of electricity
in all processing dairies; the proportion of electrical energy
associated with these services is between 3% and 5% of total site
energy (electricity and other fuels) or around 15% to 25% of site
electrical energy.

Refrigeration is used for direct process cooling after the thermal
treatment of products as well as for the preservation of finished
products. The majority of plants use refrigeration systems

employing the vapour compression cycle: (i) refrigerant is utilised as a direct product coolant through an evaporator with the primary refrigerant; and (ii) secondary refrigerants, such as chilled water or glycol, are utilised in process treatment applications.

Compressed air use is a significant consumer of electricity in all processing dairies. While in relative terms it represents a small percentage of the total energy consumed in dairies, in absolute consumption it is still significant. Compressed air is a critical secondary energy form used throughout the dairy industry for applications including pneumatic control of processes and as direct air nozzles.

## 5.3   Possibilities for optimisation

### 5.3.1   General process optimisation

#### Thermal processing (reduction of fouling)

The capacity of a production line is largely determined by the degree of fouling that occurs during production. Because proteins, minerals and even microorganisms are deposited on the walls of equipment, the equipment must be cleaned after 10 to 20 hours' run time. As a rule of thumb, the run time for production processes below 80°C is determined by the growth of microorganisms in the equipment, and at higher temperatures by the deposition of proteins and minerals. Fouling not only has consequences for the run time but also for the cleaning time, product loss and product quality. Calculations show, for instance, that in the dairy industry 30–50% of the variable production costs are the result of fouling.

In many cases the fouling is caused by changes that occur in the product itself during processing, such as denaturation of proteins and growth of bacteria. The extent of these changes is determined by the process conditions. It appears that by adjusting the temperature–time profile during heating of the product (e.g. pasteurisation, UHT) and through – often minor – changes to the equipment, fouling can be drastically reduced without making any concessions to product quality. Computer models, such as those used in the NIZO Premia simulation platform, assist in determining the optimal conditions for reducing contamination. In the case of heating equipment, evaporators and dryers, this approach has led to a reduction of fouling of

30 to 50% for certain specialised products. In the case of a new product for children, the production time between cleaning cycles may even be extended from 2 to 12 hours (de Jong et al., 2009). However, it needs to be mentioned that this concerned a very special product with severe fouling conditions.

## CIP optimisation (e.g. OptiCIP+)

Correct CIP optimisation and continued monitoring can reap significant reductions in use of chemicals, water and energy as a result of the cleaning process. Proper management of interfaces between product and flush water and process monitoring with conductivity and turbidity sensors can reduce product loss, minimise water requirements and reduce wastewater volume and effluent load. Water is required for pre-rinsing, intermediate chemical flushes and final rinsing as well as chemical makeup water. Water quality is a critical factor in attaining a successful CIP, with potable water free from chemical or bacterial contamination required for the final rinse so as not to recontaminate the system. The volume of water used and discharged from a CIP set is minimised by having a reuse system whereby final rinse water can be collected and stored for use in the next CIP. If the CIP stages are controlled by timers, the volume of water used by a CIP system can often be reduced by optimising the rinse timers, which often when set up have unnecessarily long times.

With regard to monitoring and optimisation, OptiCIP, a monitoring device to optimise CIP cleaning procedures, can be used (van Asselt et al., 2002). The system monitors the removal of organic and inorganic fouling off-line in combination with the in-line measurement of variables such as temperature, flow, conductivity and valve settings. The turbidity of the cleaning solution is a measure of the removal of organic fouling. The calcium concentration is a measure of the removal of inorganic fouling. Conductivity measurement is used for separation of the various cleaning phases and gives an indication of the concentration of the cleaning solution used. Sharp slopes between subsequent phases indicate that rinsing and cleaning phases are properly separated (van Asselt et al., 2002). The monitoring device has been applied so far for the optimisation of a large variety of process equipment (e.g. evaporators, heat exchangers and tanks).

## Reducing product loss

If the product loss of an average production line with a capacity of 200,000 tonnes of product per year and a product value of €0.5 per kg is reduced from 0.6% to 0.1% annually, this represents a saving of half a million euros per year. Apart from the product-related savings, it also has a direct effect on the amount and fouling of wastewater, since 90% of the total results from product loss. If a company does not have its own wastewater treatment (or pre-treatment), the reduction of product loss mentioned above also results in savings of more than €150,000 on the discharge costs (for example, an average of 2,750 kg of product is discharged per day and 1 kg product per day containing 10% total dry solids amounts to €60 per year in wastewater treatment costs). Even when a company has its own wastewater pre-treatment, savings will be realised through reduced consumption of water treatment additives.

A commonly occurring cause of product loss is the incorrect positioning of valves during product changeovers. This situation can result in a product that should actually have been expelled after production remaining in a tank or pipeline. During the succeeding cleaning, the product is discharged as wastewater. Improved process configuration therefore results in an immediate double benefit: less product loss and lower discharge costs (de Jong et al., 2009).

### 5.3.2  Energy use

There is potential to improve the conversion efficiencies for services in all dairies. These include optimisation of refrigeration performance and of compressed air, steam and hot water systems.

Steam is the largest secondary energy form used throughout the dairy industry for a wide variety of processes. These processes include pasteurisation, sterilisation, UHT, process air heating, evaporation, indirect product heating, direct contact injection heating, and indirect hot water generation. Refrigeration plants are a significant consumer of electricity in all processing dairies, with the proportion of electrical energy associated with these services being between 3% and 5% of total site energy (electricity and other fuels) or around 15% to 25% of site electrical energy. Refrigeration is used for direct process cooling before and after the thermal treatment of products as well as for preservation of finished products. Compressed air use is a significant consumer of electricity in all processing dairies. While in relative terms it represents a small percentage of the total energy consumed in dairies (typically 5% to 10% of site electrical requirements), in absolute consumption it is still significant. Compressed air is a critical secondary energy form used throughout the dairy industry for applications including pneumatic control of processes, direct air nozzles, pumps and motors.

## Heat recovery/reuse of heat/energy pinch

As well as delivering secondary energy services efficiently to the process, there are many opportunities in dairy processes themselves for heat recovery to reduce the overall net energy requirement. The feasibility of implementing such schemes depends on the location of the heat source with respect to the heat sink, the capital cost of the heat recovery solution and the energy savings achievable. The simplest common example in a dairy process is regenerative heating in a pasteuriser whereby the incoming cold milk is heated to a temperature approaching the final pasteurisation temperature by the returning hot milk stream. Heat recovery of up to 95% is possible, thereby minimising the net heat input to the system. This system is straightforward as the requirement for the heating step is exactly synchronous with the cooling step and the heat loads are closely matched. Increasing the regenerative efficiency from 85% to 91% for a 38,000 kg/h pasteuriser resulted in annual energy savings of 880 MWh (Korrstrom, 2001). The calculated savings for a retrofit to nine heat exchangers in a Danish dairy to increase regeneration efficiency between 4% and 6% were 2,712 MWh thermal energy and 542 MWh electrical energy per annum for an investment cost of €370,000 and a payback of 3.6 years.

For more complex situations, energy pinch analysis is a powerful process integration tool that can identify means to reduce utility costs and improve the heat efficiency of a process through a rapid understanding of the major factors influencing the heat efficiency of the process. This involves analysing the heating and cooling requirements of process streams and matching these requirements to minimise the total heat input into the process. Applications range from heat exchanger networks to complex industrial networks (Kemp, 2007). An example of heating and cooling steps that appear closely related but are out of step by about 3 hours is pasteurisation of cheese milk followed by cooling to the set temperature and then cooling of the whey from the cheese manufacture for subsequent processing. A number of cheese factories have installed stratified water storage tanks to recover heat from the whey produced during cheese making to indirectly preheat the incoming cold milk using an intermediary, thereby overcoming the constraints due to the physical layout of the plant and the operating time difference between milk pasteurisation and whey cooling. When milk pasteurisation commences, warm water at 28°C is drawn out of the top of the storage tank to preheat the cold milk. The cold water at 8°C is returned to the bottom of the tank. When whey cooling commences some 3 hours later, the tank is isolated from the process and the whey is cooled against the milk. When pasteurisation ceases, cold water from the tank is then used to cool the whey. Hot water generated is returned to the top of the tank. By the time whey production ceases, the tank is nearly full of hot water ready to supply preheating for pasteurisation, and the process is repeated. This system avoids the need for hot water at the start of cheese making and chilled water at the end of the process.

Other examples of heat recovery include recovery of heat from boiler flue gases with economisers to preheat the boiler feed water or recuperators to preheat the combustion air.

Powder production is one of the most energy-intensive dairy processes and there has been a progressive evolution in preheater, evaporator and dryer design on the energy efficiency of the manufacturing process. Numerous systems are used to recover waste heat from hot condensate from the evaporators and from the exhaust air from spray dryers. The potential for heat recovery from evaporator condensate varies with the type and efficiency of the evaporator.

Evaporators incorporating MVR technology are thermally efficient and much effort has been focused on improving the efficiency of the preheat system including staged flash/regeneration systems to minimise the amount of live steam needed for the final heat step. Many plants use the waste heat from the hot MVR condensate, which is available at 65°C, for heat recovery, particularly in the first preheating step. While these designs minimise the final condenser cooling water requirement, there always remains a cooling water load, usually taken care of by an evaporative cooling tower.

The most thermally efficient dryer configuration uses a combination of spray chamber, integrated fluid bed and external vibrated fluid bed. These designs allow more energy to be extracted from the hot air streams before exhausting to the atmosphere. Recent developments have enabled the bank of exhaust cyclones to be replaced with CIP-compatible bag filters with the advantage of reducing pressure losses in the air handling system and consequently reduced exhaust fan power consumption.

There is still substantial waste heat in the dryer exhaust. Loon (IDF, 2004) reported that, while this waste has been considered as a possible source of heat for preheating the inlet air to the dryer, in practice this is not recommended (IDF, 2004) due to the microbial contamination risk from humid, powder-containing air. GEA Process Engineering (IDF, 2004) suggest the opportunity to combine heat sources from the waste heat from the MVR condensate and the heat load from the final condenser with the opportunity to preheat the dryer inlet air (heat sink). In the example offered by GEA Process Engineering for a 2.5 tonnes-per-hour skim milk powder plant, an energy saving of 900 kWh/h is claimed, or up to 1.3 GJ/tonne powder, depending on ambient conditions and preheat treatment used. This amounts to a more than 10% reduction in specific energy consumption. This is a further example where, even though the individual process steps may already be optimised to a practical extent, application of a process integration energy pinch approach can still yield substantial energy savings. It should be noted that, for this opportunity to be realised, both the evaporator and dryer have to be running, so existing systems need to be in place during startup and shutdown, while the energy integration takes over during steady-state running.

For other examples of heat recovery and process integration the reader is referred to IDF (2005).

### Boiler strategies

Efficient supply of steam and hot water for process requirements should focus on efficiency of boiler operation, minimising losses in distribution, effective use of steam or hot water in the process and maximising condensate recovery and return to the boiler.

While the majority of steam boilers are energy efficient, there is still some potential to upgrade generation efficiency for boiler plants. The main potential for the upgrade of individual boiler efficiency is normally in one or more of the following:

- enhanced combustion efficiency;
- retrofit and refurbishment of boiler economisers;

- improved blowdown control;
- condensate recovery;
- use of variable speed drives for combustion fans.

There is still some potential to upgrade combustion efficiency for boiler plants. Minimising oxygen levels to the lowest practical level reduces the amount of excess air that is heated but not required for the combustion process. There is scope to retrofit electronic controllers to optimise combustion efficiencies for boilers. These units are available for both boilers and direct fired dryers to control oxygen and combustion chamber pressure, with options to incorporate full modulation gas valves and variable speed drive for combustion fans. Typically these can save between 2% and 4% of fuel input to boilers.

There has been an increased use of economisers on larger boilers. Most new boilers with a steam output capacity of 10 MW or more now have economisers installed on their exhaust to further recover heat by extracting heat to preheat the feed water to the boiler. Typically this reduces exhaust temperatures to approximately 120 to 130°C (typically from 220 to 240°C at boiler outlet). Economisers thus normally reduce energy input to the boiler by about 4% to 4.5%. Economisers can also be retrofitted to older boilers and/or installed on smaller (i.e. <10 MW) boilers, but because of the current relatively low cost of fuel they are normally not specified on new, smaller units nor retrofitted to older boiler plants.

Large new boilers have variable speed drives for their combustion fans instead of the more common damper control. A well set-up combustion fan allows for tighter air–fuel ratio control, delivering better combustion over the total firing rate for the boiler.

At most sites more potential savings are possible through reducing standing losses by improving boiler scheduling. Most steam boiler plant in dairies operates at average boiler load factors anywhere from 25% up to 55%. Given the reliability of new boilers there is real potential to improve load factors through reduction in standing losses by reducing the number of boilers operating to satisfy the load requirements.

There is potential to replace steam boilers with hot water units for specific applications. If the process temperature is < 90–100°C, e.g. for pasteurised milk or cheese production, low-pressure hot water can replace steam. This is more efficient due to lower flue gas heat losses with hot water temperatures. There are also lower surface heat losses from piping and process plant surfaces.

Measurements in the past have indicated that factories operating on hot water show reduction in fuel consumption of approximately 30% for pasteurised, bottled milk and 15% for cheddar cheese production, with little increase in electricity consumption compared with steam-heated plants.

While not exactly innovative, there is always some potential to save steam by increasing the level of maintenance on steam systems. Steam traps and air vents fail regularly so should be checked to ensure they are not faulty. Surveys undertaken by steam trap companies regularly identify faulty traps and vents (typically between 15% and 30% of all traps are reported as failed in surveys).

Steam condensate returns are valuable and their reuse needs to be continually assessed. Recovered condensate leads to savings in water and water treatment costs for replacing the lost condensate, savings in heat required to warm the new boiler feed water and a reduction in wastewater as less condensate enters the wastewater stream. A boiler system with 80% condensate return at 70°C can reduce fuel energy demand by 7% (IDF, 2005).

In dairies there is generally too much condensate available (combination of steam and evaporation condensates) so the use of the heat in the condensate and the value of the cleaner streams as mains water replacement need to be considered. Packaged RO units are becoming viable as potential secondary treatment devices for recovering incremental condensates.

When the pressure of saturated condensate is reduced, a portion of the liquid 'flashes' to low-pressure steam. Depending on the pressures involved, the flash steam contains approximately 10% to 40% of the energy content of the original condensate. In most cases, including boiler blowdown, condensate receivers and de-aerators, the flashing steam is vented and its energy content lost. However, a heat exchanger can be placed in the vent to recover this energy.

**Combined heat and power**

Dairy sites consume significant amounts of low-grade heat and electricity. This can provide potential opportunities to consider co-generation of heat and power for large multi-product sites. By utilising thermal energy that is always available as a result of electricity production, plants can gain an overall energy efficiency advantage of 20% to 25% compared with separate supply of steam and electricity (IDF, 2005), with consequent significant reductions in $CO_2$ emissions.

### 5.3.3  *Water use*

#### Reuse of water and possible applications

Considerations for the reuse of water include:

(1)  potential risks to food safety;

(2)  cost of water treatment including, where applicable, reduced trade waste costs compared to the cost of potable water;

(3)  potential increases in energy costs related to treatment.

In determining the potential risks associated with reusing water from either single or combined processing sources, three categories of water use have been identified:

(1)  direct contact – contact directly with food or with surfaces that come into contact with food;

(2)  indirect contact – reuse of water inside a food processing environment that is not intended for direct contact with food or food contact surfaces;

(3)  non-contact – reuse outside the food processing environment.

There can be a variety of sources of potentially reusable water within dairy manufacturing facilities. These sources and the extent to which the water can be reused (either single reuse or continual recycling) need to be identified and the following aspects considered:

(1)  hazards associated with the water source;

(2)  intended reuse of water and all impacts on food quality and safety;

(3)  whether the use is direct contact, indirect contact or non-contact;

(4)  volume of water available compared to the demand;

(5)  type of treatment process(es) to be used;

(6)  reticulation system from point of treatment to point of reuse;

(7)  potential impact on human health of non-contact uses that may produce aerosols (e.g. water used for external cleaning or irrigation).

In undertaking a water reduction assessment programme the following steps are useful:

(1) Calculate baseline water use and business activity indicator (e.g. L water/L milk).

(2) Identify water use by area and amount (data collection and analysis, possibly sub-metering).

(3) Set water-saving targets (at least 10% reduction).

(4) Identify and assess water-saving activities – cost/benefit.

(5) Complete an action plan (short-, medium- and long-term activities).

(6) Implement plan, record water use and communicate.

(7) Report annually.

In considering water reuse options, the waste minimisation hierarchy should first be applied:

• eliminate or reduce water consumption where possible;

• consider internal reuse of water (fit for purpose);

• replacement of potable water with alternative sources.

There are numerous opportunities to reduce water use in dairy processes (Prasad et al., 2005). These include:

(1) optimise water flows on pump seals, sprays, homogeniser cooling water;

(2) design and select equipment to minimise cleaning;

(3) consider product displacement air purges and consider dry cleaning before wet cleaning;

(4) optimise product changeovers to reduce cleaning;

(5) report and repair leaks immediately;

(6) optimise CIP by fine-tuning wash cycle times, in particular rinse steps;

(7) optimise crate washer operation by recirculating rinse water;

(8) use trigger-operated hoses;

(9)   optimise operation and maintenance of boilers and cooling towers;

(10)   recirculate equipment-sealing water;

(11)   install water-efficient products in amenities.

When considering reuse options, the steps should include identification of all water users and the demands of each user (sink) in terms of quality, quantity, temperature and frequency of use. The next step is to identify potential suppliers of water (also in terms of quality, quantity, temperature and frequency of use) and try to match supply and demand. When the attempted matches do not fit, process know-how must be applied to provide the best-fit solution, and the whole system must be designed to safeguard product quality and hygiene. Typical water sources in dairy processes that might be considered for reuse are condensates (steam condensates and evaporator condensates from milk or whey powder production), RO permeates from milk or whey during its processing, and final water rinses from cleaning cycles. Treatment methods include RO polishing, chemical, UV and heat treatment or combinations selected as appropriate to provide fit-for-purpose water sources, taking into account additional energy consumption and environmental impact.

Dairy powder manufacturers generally have an excess supply of condensate from the evaporators. The extent to which they can utilise this condensate depends on safeguarding the quality of the condensate. Hot and cold condensates should be separated – hot condensate used in boiler and hot water ring main and cold condensate to recovery system, storm water or waste treatment, the final destination based on conductivity and temperature parameters. Best practice low water users generally employ full CIP rinse and chemical recovery systems as CIP is usually the largest use of water on site. Comprehensive metering and monitoring systems are generally required to safeguard water reuse systems. These include flow rate, temperature, conductivity, pH and turbidity monitoring.

## Water pinch

Water pinch analysis has been adapted from energy pinch techniques and can be used to find water savings and reduce wastewater disposal in water-using networks. Water use and wastewater generation may be reduced using one or a combination of the following three principles:

(1)  Water *reuse* in operations where the level of contaminants from a previous process is acceptable. Blending with fresh water may also be done.

(2)  Wastewater may be partially *regenerated* by local treatment units before reuse.

(3)  Wastewater may be completely regenerated and *recycled*.

At its simplest, water pinch may deal with just one contaminant variable, and graphical techniques can be used. More commonly, multiple contaminants must be considered and complex algorithms are likely to be required. There are several sources of commercial software. Water pinch is still evolving and in combination with energy pinch can provide new insights into reducing the environmental footprint of industrial processes. The principle of water pinch is to match the contamination requirements between sources and sinks. It provides tools to determine water reuse target, minimum wastewater disposal, minimum fresh water consumption and regeneration targets. Even if the full analytical solution is not applied, the discipline of segregating process steps into sources and sinks, contaminant loading, minimum and maximum allowable concentrations, temperatures and flows can identify practical opportunities. Peng et al. (2008) reported the application of water pinch analysis to a large multi-product dairy processing site. Results showed an opportunity to reduce fresh water intake by 60% and correspondingly reduce effluent discharge by 65%, with a significant reduction in contaminant loading (from a base case of no water reuse). The authors commented that through logical evaluation of the water-using and contaminant streams, similar water reuse strategies could have been uncovered. However, through water pinch technology the optimal solution was obtained in a shorter time.

### 5.3.4  *Waste stream valorisation*

#### Reuse of side streams

Every production company has a waste stream that, in principle, contains (more or less) valuable components that disappear into the sewer and for which various fees must be paid. It is therefore worthwhile to subject the waste stream to a 'valorisation scan'. The purpose of this scan is to provide a complete analysis of the components and an estimate of their value. Then a study is performed to determine what technologies are required to recover the components from the waste stream and whether this is economically feasible. Examples include fresh vegetable and fruit sauces from cutting waste and the extraction of valuable proteins from meat scraps.

## The whey example

The most striking example is probably one from the dairy industry. In recent years whey originating from cheese production (see Figure 5.3) has been increasingly used as an ingredient in high-value products including animal feed and bioactive proteins in sports drinks. The waste streams that still remain contain minerals such as calcium, phosphorus and potassium. Research is currently under way to determine

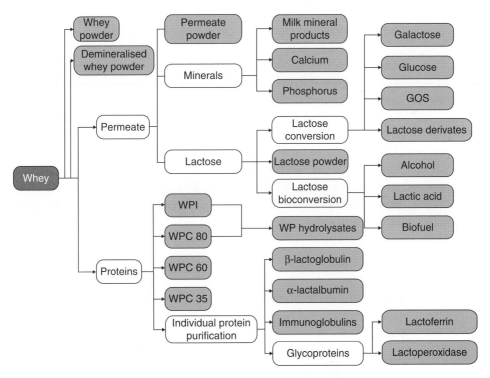

**Figure 5.3   Whey fractionation.**

whether new concepts for utilising these components are economically feasible.

## 5.4    Revisiting dairy processing: breakthrough technologies

### 5.4.1    Model-based dairy production

A further development is the on-line application of computer models for real-time monitoring and optimisation of processes. An example of such a system is NIZO PREMIC. NIZO PREMIC can be installed on top of the existing process control system such as, for example, the DCS/SCADA system of Honeywell. PREMIC is fed with data concerning the raw materials (such as composition), current status of the process (via DCS/SCADA) and resulting end-products. PREMIC uses calculation modules from NIZO Premia. Depending on the type of unit operations the relevant Premia module can be loaded into PREMIC. The interface of PREMIC can be accessed via the DCS/SCADA system.

### Communication between process and PREMIC

There is continuous communication between process, DCS/SCADA and PREMIC. Current temperatures, flows, etc. are transmitted to PREMIC via the DCS/SCADA system. Based on this information PREMIC calculates set point values necessary to ensure a high product quality and a minimum of fouling. These set points are fed back to the DCS/SCADA system. It is possible to automatically replace current set points with set points advised by PREMIC, but it is also possible to only show advised set points and let the operator decide whether to replace the current set point.

### Proof of principle

The model-based control system developed has been validated on the pilot-scale heat exchangers of NIZO Food Research. Two pasteurisation experiments were carried out with a 20% (w/w) solution of low-heated milk powder.

In the first experiment traditional process control without PREMIC was applied and the heating medium was adjusted by the DCS/SCADA system in order to keep the pasteurisation temperature constant. Pasteurisation temperature was 120°C and holding time was 15s. After 3h of production the process was stopped and fouling of the apparatus was monitored visually. During production, temperatures at different points in the process as well as the product flow were measured continually. At the end of the production run these data were used to calculate the concentration of β-lactoglobulin in the end-product at different points during the production run.

In a second experiment the same whey and pasteuriser were used, but set points were controlled with PREMIC in order to minimise fouling and achieve a more constant product quality in terms of denaturation of β-lactoglobulin. Again the process was stopped after 3h and fouling was inspected visually.

The latest development in this field is OptiCIP+. By using a CIP model combined with a model-based process control system (PREMIC), a system is now available to continuously monitor, optimise and control the CIP procedure. Based on the actual production data, the type of product and two in-line sensors (turbidity and $Ca^{2+}$), the OptiCIP+ system determines the optimal CIP procedure. During cleaning all CIP-related data are linked with the production and product composition data and stored in a database file.

After production, this database file is searched and, depending on the process conditions, product composition and degree of fouling, an initial CIP procedure is loaded into the system. By making use of the sensor values, the CIP procedure will continuously be optimised during cleaning.

From an evaluation study in the industry it turned out that CIP procedures are based on worst-case scenarios and there is a potential for 30% saving in cleaning efficiency and thus reduction of carbon footprint (Van Asselt et al., 2005).

### 5.4.2   High solids evaporation and drying

As shown in sections 5.2.3 and 5.2.4, evaporation uses far less energy compared with drying. The most pragmatic approach therefore would be to remove more water during evaporation, resulting in higher dry solids content before drying. Apart from the reduced energy use this also implies a higher capacity for the spray dryer. The effect of higher dry solids content on the spray drying capacity is shown in Figure 5.4.

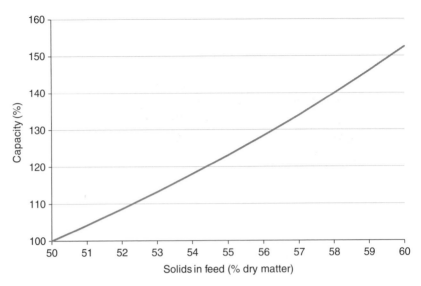

**Figure 5.4 Spray dryer capacity as a function of the dry matter concentration of the concentrate after evaporation, assuming that the capacity at 50% dry matter for the feed equals 100%.** Reproduced from Fox et al. (2010), with permission of Russell Publishing Ltd.

# References

Van Asselt, A.J., van Houwelingen, G., te Giffel, M.C. (2002) Monitoring system for improving cleaning efficiency of cleaning-in-place processes in dairy environments. *Food and Bioproducts Processing*, 80: 276–280.

Bargeman (2003). Separation technologies to produce dairy ingredients. In Smit, G. (Eds.), *Dairy processing: improving quality.* Cambridge: Woodhead Publishing, pp. 366–390.

Carbon Trust (2011) www.carbontrust.co.uk/cut-carbon-reduce-costs/calculate/carbon-footprinting/Pages/conversion-factors.aspx.

Dairy Australia (2009) *Australian Dairy Manufacturing Industry Sustainability Report 2007/2008.*

Dairy UK (2009) *Sustainability Report 2009.*

de Jong, P. van Asselt, A., Fox, M., Akkerman, C (2009) Five measures for sustainable, financially sound processing. *New Food*, 4: 22–23.

FAO (2010) *Greenhouse gas emissions from the dairy sector: a life cycle assessment.*

Flapper, J. (2009) Environmental impact analysis and benchmarking in the dairy processing industry. Thesis, University of Groningen, Groningen.

Fonterra (2011) www.fonterra.com/wps/wcm/connect/fonterracom/fonterra.com/Our+Business/Sustainability/.

Fox, M., Akkerman, C., Straatsma, H., de Jong, P. (2010) Energy reduction by high dry matter concentration and drying. *New Food*, 2 (12 May): 60–63.

International Dairy Federation (2004) Advances in fractionation and separation: processes for novel dairy applications. *Bulletin of the International Dairy Federation*, no. 389. Brussels: IDF.

International Dairy Federation (2005) Energy use in dairy processing. *Bulletin of the International Dairy Federation*, no. 401, 77. Brussels: IDF.

International Dairy Federation (2009) Environmental/ecological impact of the dairy sector: literature review on dairy products for an inventory of key issues. List of environmental initiativess and influences on the dairy sector. *Bulletin of the International Dairy Federation*, no. 436. Brussels: IDF.

Kemp, I.C. (2007) Key concept of pinch analysis. In Kemp, I.C. (Ed.), *Pinch Analysis and Process Integration: a user guide on process integration for the efficient use of energy.* Oxford: Elsevier.

Korsstrom, E., Lampi, M. (2001) *Best available techniques for the Nordic dairy industry,* www.norden.org/en/publications/publications/2001-586.

Oldenhof, S. (2004) *Uitgebreide Energiestudie Zuivelindustrie.* Senter Novem BV.

Peng, S.F., Farid, M. Ho, I, Mahdi, K.A. (2008) Water and waste water minimisation in dairy plants using water pinch technology. *Journal of International Environmental Application and Science*, 3(1): 43–50.

Prasad, P., Pagan, R., Kauter, M., Price, N., Crittenden, P. (2005) *Ecoefficiency for Australian Dairy Processors*; Fact sheet 1 Water Management.

Ridoutt, B.G., Williams, S.R.O, Baud, S., Faval, S., Marks, N. (2010) The water footprint of dairy products: case study involving skim milk powder. *Journal of Dairy Science*, 93(11): 5114–5117.

Rowlands, W. (2011) Innovations in homogenizer and separator technology for the modern dairy plant. *Journal of Dairy Science*, 94(Suppl. 1): 479.

US Dairy (2010) *U.S. Dairy Sustainability Commitment Progress Report*.

van den Berg, H., Jansen, L.A., Karsmakers, J.P.H (1983) *Sectoronderzoek Energiebesparing in de Zuivelindustrie*, Mededeling M17. Ede, the Netherlands: NIZO, pp. 225.

Verdurmen, R.E.M., de Jong, P. (2000) Optimising product quality and process control for powdered dairy products. In Smit, G. (Ed.), *Dairy processing: improving quality*. Cambridge, UK: Woodhead Publishing, pp. 333–365.

Vissers, M.M.M., de Jong, P., de Wolff, J. (2002) New evaporator design results in 70% reduction of energy consumption. *Voedingsmiddelentechnologie*, 20–22.

# 6

# The role of packaging in a sustainable dairy chain

Erika Mink

Tetra Pak International, Brussels, Belgium

**Abstract:** The sustainable performance of packaging has been a long-time focus for business, regulators and non-governmental organisations. The role of packaging in supporting the dairy sector's drive to meet its sustainability challenges – which for good reason have so far mainly focused on farming – is now receiving closer attention. There is growing pressure for information and evidence to demonstrate the environmental sustainability of all sectors including dairy. The worldwide dairy industry responded in 2009 with its Global Agenda for Action on Climate Change. The Agenda has emerged within a context where a growing number of governments are adopting greenhouse gas reduction measures, and legislation on producer responsibility for packaging waste is spreading globally. Retailers, facing mounting consumer expectations and closely monitored by NGOs, are active in driving sustainability requirements throughout their supply chains and have put a strong emphasis on packaging in achieving this. Leading consumer goods companies and retailers have together developed a set of global metrics for packaging sustainability, likely to become a worldwide reference.

*Sustainable Dairy Production*, First Edition. Edited by Peter de Jong.
© 2013 John Wiley & Sons, Ltd. Published 2013 by John Wiley & Sons, Ltd.

Looking further ahead, global food security will make the prevention of food losses, particularly those linked to distribution and consumption, a major priority for packaging, and also create a sustainability challenge whose environmental dimensions may increasingly share prominence with its economic and social imperatives.

**Keywords:** carbon footprint, environmental impact, packaging, recycling

## 6.1    Introduction

The essential role of packaging is to protect food and prevent food losses, avoiding wastage of resources and thereby contributing to the goal of living within the limits of the planet's resources. Safe and efficient milk distribution from the dairy plant to the consumer's table is packaging's most important economic, societal and environmental function. It also performs other roles throughout the life cycle of dairy products (Figure 6.1). For example, in addition to supporting efficient product transport and handling over the entire supply chain, packaging creates consumer convenience in product use, matches the variety of consumer demand with different portion sizes, helps differentiate and promote products, and carries consumer information on, for example, nutrition, health and packaging disposal (Consumer Goods Forum, n.d.).

In a world constrained by the availability of resources, these packaging functionalities have to be achieved with minimum amounts of resources. Major reductions in packaging weight have been achieved over the years and have been a driving force for eco-innovation. But weight reduction has its limits. There is an optimum quantity of packaging material needed to ensure product protection (Figure 6.2). As illustrated, under-packaging can cause product losses leading to far higher negative environmental and economic impacts than those attributable to over-packaging.

The risk of under-packaging has been widely acknowledged:

The industry has a responsibility to review the packaging it uses and to ensure that any negative impact arising from its production or disposal is minimised. But this analysis of

**Figure 6.1    Packaging in the dairy value chain.** *Source*: Tetra Pak (2011).

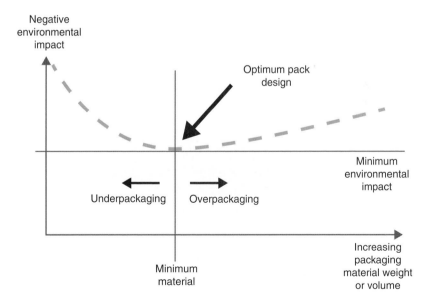

**Figure 6.2    Optimum packaging: environmental impacts.**
*Source*: Innventia AB (2009) *Packaging in the Sustainability Agenda: A Guide for Corporate Decision Makers*. Reproduced by permission of Innventia, ECR Europe and EUROPEN.

impacts must be done in the round. It must include the impact of product losses that may result from the use of too little packaging as well the impacts of using too much. (Consumer Goods Forum, 2010)

Optimised packaging, which is adapted to specific distribution channels and local conditions, refers to packaging systems (Figure 6.3) which include:

- the packaging around the product itself – primary packaging;
- grouped packaging holding together different sales units of packed products – secondary packaging;
- transport packaging used to distribute sales units from producers to retailers – tertiary packaging.

Nevertheless, even optimised packaging consumes resources and generates environmental impacts during material extraction, production, distribution and its final disposal. Packaging is estimated variously to contribute 5% to 7% to the overall climate impact of milk (see section 6.3.1) depending on the farming systems, the types of packaging used and the waste management options available. An impact reduction strategy for the dairy

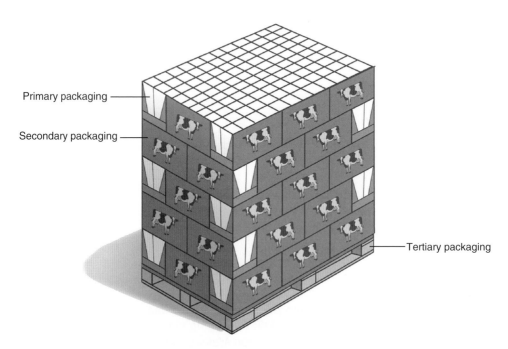

**Figure 6.3   Example of a packaging system.** *Source*: Tetra Pak (2011).

chain as a whole can be optimised by including the impacts linked to packaging in the scope of reduction.

Examination of some of the chief factors influencing these efforts at impact reduction is set out in the following sections of this chapter. Section 6.2 considers how consumer, business and regulatory pressures are driving increased expectations for the sustainability of packaging generally, and suggests some specific consequences for dairy packaging. Packaging's role in the environmental performance of dairy, as it seeks to implement a vision of sustainability for the sector through the Global Dairy Agenda, is outlined in section 6.3. The emerging alignment of packaging requirements worldwide, discussed in section 6.4, provides the dairy industry with a new and coherent framework for integrating packaging's contribution into its sustainability efforts in food distribution globally. Using the example of Tetra Pak, section 6.5 summarises one company's response to sustainability expectations with specific reference to the dairy chain. The chapter finishes with a brief perspective (section 6.6) which suggests that recent developments – e.g. concerns about food security and high food prices – may be creating a new type of sustainability challenge for the packaging and dairy sectors in the coming years. Beyond environmental pressures, sustainability's economic and social pillars may gain added significance.

## 6.2    Packaging sustainability: a growing market expectation

Packaging in general has been in the sights of environment groups, citizens and legislators around the world for many years. An initial peak was reached in the early 1990s in North America and Europe, with legislative measures (e.g. the EU Packaging and Packaging Waste Directive) spurred on by NGOs. Since then, stakeholder pressures have spread worldwide to include in particular the global retail sector, and tended to focus on the packaging used for everyday consumer products such as food and drink. Industry is expected to deliver value-for-money products in packaging adapted to ongoing changes in demography and people's lifestyles.

Global consumer trends (Tetra Pak, 2009) suggest that dairy products are responding to changing lifestyles. Even though consumption patterns and habits differ between countries and cultures – for example, in some parts of the world 'milk' is a synonym for food or a staple element of people's nutrition – a point in common is that dairy suppliers have to meet the needs of increasingly mobile consumers with little time for frequent shopping, high demands for convenience, and 'on-the-go' eating habits (GfK Roper Consulting, 2009). At the same time, modern lifestyles are seen as wasteful – with packaging the icon *par excellence* of wastefulness. Accordingly consumers, who expect as a matter of course that packaging meets the various demands of modern lifestyles, tend, as citizens, to see their involvement in recycling as a means of preventing valuable resources being wasted.

### 6.2.1   Consumer expectations

Recycling is the no. 1 environmental expectation of packaging among global consumers (Petrie, 2011). Research shows that consumers in the UK, France, Brazil, the US, Japan, Russia, China, Germany and India mention recyclability (37% on average) and litter (22%) as top-of-mind concerns when asked about the environmental impact of milk and juice packaging. Around 45% say they would even be willing to switch brands on the grounds of more environmentally friendly packaging, this view being most strongly held in India and China. When asked to rate the overall factors influencing their choice of milk container, environment ranks third (but with growing importance since 2005) after convenience and food safety (GlobeScan Inc., 2009; Young & Ciummo, 2009).

Meeting consumer expectations on recycling is therefore crucial, yet at the same time it is becoming more challenging. Some good progress has been made in developed markets, with the creation and expansion of collection schemes for used packaging enabling an increase in recycling rates. However, the situation is more challenging in emerging markets. They are characterised by continued growth in product demand and thus in packaging, but also by insufficiencies in municipal waste management that hinder the needed rapid growths in recycling.

It is expected that recycling will remain the dominant consumer expectation overall of packaging worldwide, but there are others too. Glass and renewable packaging materials, for example, generally receive more favourable environmental perceptions than packaging composed of other materials. Green consumers, for their part, are pushing for greater transparency on life cycle impacts, including more in-depth information on the climate impact of packaging.

## 6.2.2  *Worldwide legislative pressures*

Governments worldwide have legislated on packaging in order to manage and reduce household waste and littering and also to shift the cost of managing used packaging from public entities to producers. A number of countries have regulated packaging since the early 1970s, but Germany's Packaging Ordinance in 1991 (German Federal Ministry, 1991) was the first comprehensive law requiring extended producer responsibility (EPR) – whereby consumer goods companies and producers of packaging become financially liable for managing packaging waste. EPR-based legislation for packaging began to spread rapidly within Europe and beyond after the EU adopted its Directive on Packaging and Packaging Waste in 1994. At the same time, packaging and recycling regulations were becoming more common in Asia.

Today, products sold in the Americas (e.g. in the US, Canada, Brazil and Mexico), in Asia/Pacific (Australia, China, Japan, Taiwan, South Korea and India), in Africa and the Middle East (South Africa, Israel) and Europe (EU, Turkey, Ukraine) are all subject to some form of environmental packaging requirements. The list is continually growing as governments look to address waste and resource management issues and also seek ways to fund their municipal waste collection systems.

The aim, scope and requirements of packaging environment regulations vary widely from country to country, but there are convergent trends. In addition to EPR, the most significant requirements include mandatory recycling targets, design standards, material efficiency requirements and packaging taxes. Under many producer responsibility policies designed to help meet recycling targets, producers of packaged goods are obliged to set up or to finance the collection and recycling of used packaging. Both waste collection fees and taxes have traditionally been based on the amount and on the type of packaging materials put on the market. Now they are also increasingly linked to the overall environmental performance of the package, and here climate impact is often used as the lead indicator for packaging's life cycle performance.

Legal incentives may in addition be considered to encourage the use of packaging with favourable environmental features, for example, packaging:

- with low environmental impacts along the life cycle;

- that is recyclable and recycled;

- that demonstrates low climate impact.

### Beyond recycling targets

The German Packaging Ordinance (German Federal Ministry, 2005) already goes well beyond recycling targets. It promotes the use of 'ecologically advantageous packaging' (e.g. refillable bottles, beverage cartons) which have been identified through a life cycle assessment (LCA). French legislation is moving even further. A law passed in 2010 (Government of France, 2010) stipulates that as of 2012 environmental information (including impacts on climate) has to be made available at points of sale for products and their packaging. Driven by resource concerns, China's Recycling Economy Promotion Law (People's Republic of China, 2009), which came into force in 2009, decrees a recycling planning system and includes a system to audit progress in the recycling economy. For its part, the EU's emerging resource policy is likely to highlight waste prevention and promote the use of recycled materials.

### Outlook

Legislative pressures on packaging are expected to persist in countries with legislation in place and to increase in other geographies. As many as 40% of decision-makers and policy influencers (in governments, NGOs, business and media) in the UK, France, Brazil, the US, the EU institutions, Japan and Russia see 'environment and energy' as the most important factor shaping the future for beverage and food packaging. Decision-makers in China, Germany and India see it as the second most important factor after 'health and safety' (GlobeScan Inc., 2009).

The evidence suggests that the future will be marked by a growing range of regulatory demands for environmental performance and an increasing financial burden on packaging producers and users as they seek to meet them.

### 6.2.3   Packaging: a priority focus for NGOs

Packaging is often perceived as a resource-intensive way of distributing products in a resource-strapped world and one which ends up consigning tonnes of packaging waste every day to landfill. By their actions and campaigns over the years, NGOs and environmental groups have come to be seen by consumers as the most credible source of information on packaging and its environmental impact, and thus they can significantly influence consumer behaviour in the marketplace. Today they continue to raise consumer awareness by keeping the packaging issue in the public eye. As the WWF states:

> Packaging deserves more attention than most of us would think. While packaging performs the key functions of protecting a product and providing information for buyers, what we tend not to think about is what happens to it later. Why, it has to be disposed of, of course. ... When it comes to disposal, most packaging is not recycled, but dumped in landfills or burnt. (WWF, n.d.[1])

At the same time the relationship between NGOs and industry is gradually changing. While NGOs continue to push for less packaging and in favour of reusable packaging, there is also a greater emphasis on dialogue with business. They engage in discussions with industry and press for commitments from companies and their supply chains to tackle packaging waste and to offer more sustainable products in sustainable packaging. NGOs are also pushing retailers to enhance the sustainability of their product offerings to consumers by using 'choice editing' (European Environmental Bureau, 2009). Their aim is to seek commitments from retailers only to put on their shelves packaged products meeting environmental criteria, so that consumers can choose from a range of items all of which are by definition environmentally sustainable.

In recent years several NGOs have extended the scope of the environmental information sought, to include the life cycle performance of packaging and products. NGOs have built their own expertise in LCAs and indeed sometimes act as peer-reviewers for scientific studies commissioned by governments or companies.[2] Many NGOs, both global and local, have come to work in partnerships with producers and retailers to help promote recycling and sustainable materials and products. In addition, they increasingly use market mechanisms to drive their agendas and, based on their high credibility with consumers, they can do so effectively. For example, in 2010 the German Nature and Biodiversity Conservation Union (NABU)

---

[1]  wwf.panda.org/about_our_earth/teacher_resources/project_ideas/waste_management/ (accessed 17 February 2011).

[2]  For example, the critical review committee for an LCA study carried out in France (see section 6.3) included an expert from WWF.

launched a 'green shopping basket award'. NABU awards retailers who help to promote sustainable consumption by offering organic and eco-labelled products, and also for their consumer information and product offerings in 'ecologically advantageous' packaging (NABU, 2010).

### 6.2.4   Retailers: the gateway to sustainable markets

Chiefly because of their direct consumer interface, retailers, in particular in the US and Europe, have come to assume a pivotal role in responding to the sustainability requirements faced by consumer goods in general – and, more specifically, by food and drink products and by the packaging needed to get them into the hands of consumers. Its tangible and pervasive importance – some form of packaging is required by virtually all retail consumer goods – means that packaging has been and remains a priority for retailers in their efforts to contribute to the creation and spread of sustainable markets. Market leaders such as Walmart, Carrefour and Tesco have been in the vanguard of these efforts. They are using their key position in the consumer product supply chain – at the gateway to the consumer market – to require from all their suppliers a contribution to the greening of the packaged products put on their shelves. Walmart's 2006 packaging scorecard programme, for example, requires suppliers to provide data enabling comparison of their performance with their competitors' on indicators such as recycled content and GHG emissions. Tesco has set a packaging reduction target, and Carrefour has sought to optimise packaging weights starting with its own-brand products.

The increasingly powerful role in sustainability played by retailers is not just a result of consumer pressure and of retailers' strategic position in meeting it. They have also benefited from added public legitimacy. Policy-makers, notably at the EU, have deliberately devolved onto retailers a quasi-regulatory role in implementing Europe's political commitment to create a sustainable economy. The programme adopted in 2009 by Europe's Retail Forum, and endorsed by the EU, shows best practice by retailers in a number of sustainability areas.[3] Retail's legitimacy in this role has been strengthened by other

---

[3]   http://ec.europa.eu/environment/industry/retail/.

developments. These include the growing practice of making public and verifiable commitments about the environmental performance of suppliers to retailers, including suppliers of packaging.

Retailers are thus assuming a new leadership role on sustainability performance, as expectations of consumers, government and NGOs converge on them. They are responding both with visions – e.g. Tesco's of a zero carbon business by 2050 – and with increasingly rigorous purchasing requirements. This leading role is further confirmed by the recent preparation of global packaging sustainability metrics under the auspices of the Consumer Goods Forum, a worldwide network of retailers and consumer goods manufacturers (see section 6.4.1). In the light of these developments, the focus of retailer sustainability requirements on packaging seems certain to intensify in the years to come.

### 6.2.5  *Some consequences for dairy packaging*

Sections 6.2.1, 6.2.2, 6.2.3 and 6.2.4 have outlined the range of challenges confronting the environmental performance of packaging. In summary:

- growing consumer expectations focused primarily on recycling;

- expanding legal requirements, particularly on EPR;

- persistent and more detailed demands for information from NGOs;

- more rigorous requirements by retailers.

Faced with these challenges, the dairy sector and its packaging suppliers will need to consider action on a number of fronts, for example:

- how best to support the setting up of the collection systems needed to meet expectations and targets on recycling;

- how to build capacities and markets for a wider range of dairy packaging materials;

- in particular, how best to use the global packaging sustainability metrics for dairy packaging systems (see section 6.4).

## 6.3    Packaging's contribution to dairy sustainability

Since 1987 when the Brundtland Commission defined sustainability as 'development that meets the needs of today, without compromising the ability of future generations to meet their own needs' (Brundtland Commission, 1987), governments and forward-looking businesses have been seeking to relate the Brundtland definition to their respective roles and responsibilities. In 2006, for example, we find the EU's Council of Ministers translating it into 'People, Planet and Profit' (Council of the European Union, 2006). For its part, the vision of the World Business Council for Sustainable Development, as expressed in 2010, is that 'in 2050, around 9 billion people live well, and within the limits of the planet' (WBCSD, 2010). In the meantime, many individual sectors and companies have sought and are still seeking to work out how in practice to adapt and apply these broad visions to their business strategies going forward – not least the dairy chain, which states the following:

> The dairy supply chain is committed to providing consumers with the nutritious dairy products they want, in a way that is economically viable, environmentally sound and socially responsible. (Global Dairy Agenda for Action, 2009)

Applying in practice the environmental dimension of dairy's sustainability vision, its most relevant challenges include climate (reducing GHG emissions), sustainable use of resources and preventing waste over the product life cycle. In this, the contribution made to the sector's environmental sustainability by packaging – vital in getting dairy products to market – is not to be ignored.

### 6.3.1    Packaging in the analysis of the dairy chain's environmental impacts

Packaging's contribution to the overall environmental impact of the dairy chain is the subject of ongoing analysis. At the same time, the traditional focus of environment research and enquiry has been on farm-related emissions which represent dairy's more obvious and major impacts.

The most conclusive environmental data available for the dairy chain as a whole are for climate impact. These data generally distinguish between two phases in the dairy cycle – cradle-to-farm-gate emissions, on the one hand, and post-farm-gate emissions, on the other. The emissions ratio between the two phases varies, partly for methodological reasons, in particular because of inconsistent system boundaries applied to post-farm-gate emissions, the phase where packaging impacts occur.

Research on post-farm-gate emissions, for example, often excludes impacts from retail activities, packaging waste disposal and product losses along this part of the dairy chain. All three of these could be significant in reducing impacts in the post-farm-gate phase, even if they are unlikely to alter the basic distribution of impacts between the dairy chain's two phases. So, while cradle-to-farm-gate emissions are expected to remain the most important single stage in the life cycle on which to focus for impact reduction, the post-farm-gate phase will gain importance. It is a phase with 'low hanging fruits' in terms of impact reduction. Here product waste prevention and packaging may be underestimated in their potential to contribute to the overall reduction of GHG in the dairy chain.

A global perspective on dairy chain impacts is provided by an FAO report on GHG emissions (FAO, 2010). It concludes that a total of 2.4 kg $CO_2$-eq per kg milk (global average 2007) are emitted. Cradle-to-farm-gate emissions have by far the highest proportion of this impact, contributing on average 78% to 83% to total dairy GHG emissions in industrialised countries, and significantly more in developing markets (due to differences in farming systems and distribution). Post-farm-gate emissions (excluding final disposal of packaging) reach up to 0.23 kg $CO_2$-eq per kg milk. For post-farm-gate activities, the FAO recommends, among other measures, to choose packaging material with lower production- and disposal-related GHG emissions (FAO, 2010).

The FAO also addresses the question of GHG emissions related to packaging material production for dairy products across the world's regions. For milk, the packaging material types assessed were cartons (gable top and aseptic brick-shaped), plastic pouches and high density polyethylene (HDPE) bottles. Region- and country-specific GHG emissions were estimated by combining two metrics – average energy use for packaging per kilogram of product, and emission factors per unit of energy used. Calculations refer to the packaging container most frequently used in the region (Table 6.1).

Meanwhile, further studies provide relevant data on post-farm-gate GHG emissions. In the US, post-farm emissions represent about 20% of the overall dairy chain total impact. Of this 20%, packaging contributes up to one-third, thus representing 7% of the GHG emissions of the entire chain. Further, the report shows

**Table 6.1** Variations of packaging material impacts between regions.

| Region | Packaging alternative | GHG emissions per packaging (g $CO_2$-eq per litre milk) |
| --- | --- | --- |
| Central and South America | Carton brick, 1 litre | 56 |
| Former Soviet Union | Plastic pillow pouch 1 litre | 20 |
| Greater China | Plastic pillow pouch 0.25 litre | 52 |
| USA, Canada and Mexico | HDPE bottle, 1/2 gallon | 91 |
| Northeast Asia and Oceania | Carton gable top, 1 litre, chilled | 38 |
| Sub-Saharan Africa | Carton brick, 1 litre | 63 |
| EU-27 | Carton brick, 1 litre | 59 |
| Southern Asia, incl. Mediterranean Africa | Plastic pillow pouch, 1 litre | 23 |

*Source*: FAO (2010) *Greenhouse gas emissions from the dairy sector: a life cycle assessment*, Food and Agriculture Organization of the United Nations, April 2010. www.fao.org/docrep/012/k7930e/k7930e00.pdf. Reproduced with permission from the Food and Agricultural Organization of the United Nations.

that packaging material production represents the greatest source of emissions and accounts for approximately 65% of total GHG emissions associated with packaging (Figure 6.4).

A similar ratio between cradle-to-farm-gate and post-farm-gate emissions can be found for Europe but no conclusive results have been forthcoming on the contribution of packaging to the post-farm-gate emissions (Sevenster & de Jong, 2009).

Some further indications, however, emerge from a pilot project carried out by Carrefour and Tetra Pak for UHT milk in France (Carrefour & Tetra Pak, 2010).[4] Results show that the

---

[4] Carrefour & Tetra Pak, October 2010, internal report, *Global Packaging Project pilot project*. Note: this project was not specifically undertaken to measure the climate impact of milk and milk packages, but rather to test a set of metrics compiled by the Consumer Goods Forum for worldwide application; see section 6.4.

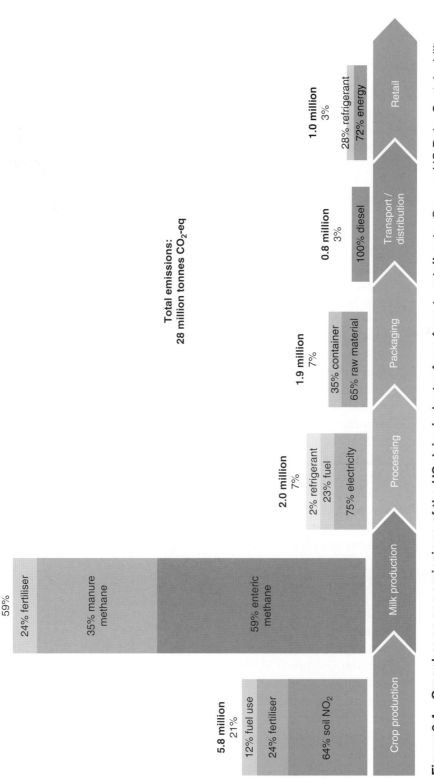

**Figure 6.4  Greenhouse gas emissions of the US dairy industry from farm to retail gate.** *Source*: US Dairy Sustainability Initiative (2008) *A roadmap to reduce greenhouse gas emissions and increase business value*, December 2008, Innovation Center for US Dairy.

16.5 million
59%

24% fertiliser

35% manure methane

59% enteric methane

Total emissions:
28 million tonnes $CO_2$-eq

2.0 million
7%

2% refrigerant
23% fuel
75% electricity

1.9 million
7%

35% container
65% raw material

0.8 million
3%

100% diesel

1.0 million
3%

28% refrigerant
72% energy

5.8 million
21%

12% fuel use
24% fertiliser
64% soil $NO_2$

Crop production

Milk production

Processing

Packaging

Transport / distribution

Retail

packaging system (using 1-litre packs) contributes 5.6% $CO_2$-eq to the overall GHG emissions, a figure that includes impacts from the entire packaging system – primary, secondary and tertiary (as explained in section 6.1). Similar relative distribution between cradle-to-farm-gate and post-farm-gate impacts was found for water, namely, 62% of freshwater consumption is attributed to the product, 16% to the packaging system and 22% to other life cycle stages (processing, distribution and storage).[5]

### 6.3.2  *Packaging's life-cycle environmental performance*

Decades of stakeholder focus on packaging and the environment have led to numerous life cycle assessments being conducted both by governments and industry on the environmental impacts of liquid food and beverage packaging systems. LCAs have been used to inform public policy and legislation, to drive supply chain improvements, to provide data for new packaging developments, and to contribute to the fact base for both business-to-business and business-to-consumer environmental communications.

Life cycle assessment uses a series of standardised environmental impact indicators, including GHG emissions. As the importance of tackling climate change is widely recognised and as GHG emissions identified in packaging LCAs tend to correlate with a number of other impact categories, climate impact has emerged as a lead indicator.

Below are three examples of peer-reviewed LCA studies comparing packaging for milk, covering impacts during the following phases: raw material production for the packaging system manufacture of the packaging, filling the product into the package, distribution of the packaged milk to retail outlets, and the end-of-life (post-consumer) phase in Europe. Web links are provided to enable reference to the full studies.

The studies, all relatively recent, look at the comparative environmental performance of different types of representative packaging used in three geographies for milk products. In Sweden (Figure 6.5), the comparison is between cartons, PET and HDPE bottles for packing pasteurised milk; in France (Figure 6.6) an HDPE bottle and cartons for packing UHT milk are compared; and in Germany (Figure 6.7) cartons and a PET bottle for packing pasteurised milk-mix drinks are compared. In

---

[5]  Oliver Muller, Price Waterhouse Coopers, October 2010, Carrefour & Tetra Pak, *Global Packaging Project pilot project*, Summary Report, Paris.

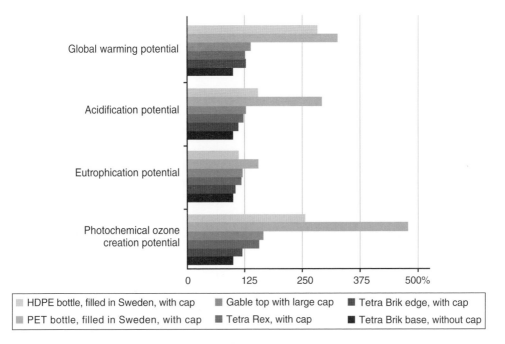

**Figure 6.5  Example 1: Comparative LCA study on packaging (1.0 L) for pasteurised milk in Sweden.** *Source*: Kristian Jelse, Elin Eriksson and Elin Einarson, Swedish Environmental Research Institute (IVL), 25 August 2009, Life Cycle Assessment of consumer packaging for liquid food – LCA of Tetra Pak and alternative packaging on the Nordic market; www.tetrapak.com/Document%20Bank/environment/climate/LCA%20Nordics%20Final%20report%202009-08-25.pdf; www.tetrapak.com/environment/climate_change/co2footprint/lca/bev_containers/pages/default.aspx.

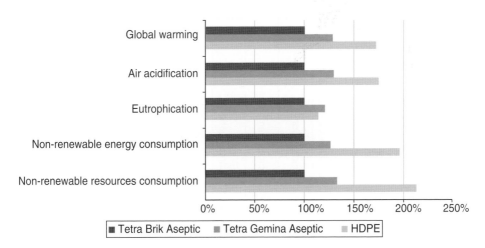

**Figure 6.6  Example 2: Comparative LCA study on packaging (1.0 L) for pasteurised milk in France.** *Source*: Bio Intelligence Service (2008) Comparative Life Cycle Assessment of Tetra Pak packaging. Synthesis www.tetrapak.com/se/Documents/TetraPak_ACV_Synthese_UK_vf_2007%5B1%5D.pdf or www.tetrapak.com/fr/environnement/co2%20footprint/choix_durenouvelable/Pages/default.aspx. Reproduced by permission of Bio Intelligence Service.

| Environmental improtance | Impact category | Beverage cartons[1] in % worse than PET bottles[2] | Beverage cartons[1] in % better than PET bottles[2] |
|---|---|---|---|
| **Very high** | Global warming potential | | 199% |
| | Terrestrial eutrophication | | 29% |
| **High** | Resource consumption (fossil) | | 217% |
| | Acidification | | 53% |
| | Summer smog | | 18% |
| **Medium** | Aquatic eutrophication | 90% | |
| | Space requirements (forest) | 1836% | |
| **No classification** | Energy consumption (total) | | 63% |

1) 0.5 L beverage cartons without aluminium, with closure, for refrigerated distribution of milk/ milk-mix drinks.
2) 0.5 L PET monolayer bottles without barrier (20 g).

**Figure 6.7  Example 3: Comparison of packages for milk-mix drinks (0.5 L) in Germany.** *Source:* Fachverband Kartonverpackungen für flüssige Nahrungsmittel e.V. (FKN) based on IFEU 2006, www.getraenkekarton.de/01_seiten/page. php?id=118, English summary: FKN, 2007, Life-cycle-assessment – Beverage cartons under test, page 21, www.tetrapak.com/ Document%20Bank/environment/climate/FKN_LCAs_publication_2007.pdf. Reproduced with permission from FKN.

all three, the measurements are made against a set of environmental indicators; the packaging system scoring best on the
climate impact (or 'global warming potential') tends also to
score well on other indicators.[6]

Further information on comparative life cycle assessments is
available. A comprehensive evaluation covering 22 comparative
and peer-reviewed studies carried out in Europe for milk and
juice packaging shows similar results (von Falkenstein et al., 2010).

### 6.3.3   *Driving packaging's contribution to environmental impact reductions*

For the dairy sector focusing on reducing environmental impact
(and using GHG as a lead indicator) along its chain, choosing
packaging which supports this goal becomes a significant step.
As a result of LCAs, environmental impacts of packaging systems are now well understood. The main share of packaging life
cycle impacts is generated by the packaging material. As shown
in the three examples above, different materials have different
impacts. The fewer resources used to produce a package and the
higher the share of renewable materials and energy used in this
process, the lower are the impacts.

A second significant part of the overall impact is generated
after the milk has been consumed and householders dispose of
the empty package. These impacts depend on available waste
management options (recycling, energy recovery, landfill), and
they vary greatly from country to country. The higher the recycling and recovery rate, the lower the end-of-life impacts.

A third area generating life cycle impacts is the efficiency of
the packing or filling process, and here the consumption of
energy and water and the production waste generated are three
determining factors.

As illustrated in Figure 6.8, two-thirds of $CO_2$-eq generated
from a 1-litre Tetra Brik Aseptic package is attributed to its materials. Figure 6.8 also shows the impact of different materials.
Aluminium, only 4% of the package's weight, contributes nearly
half the $CO_2$-eq impact of the materials. Paperboard, representing

---

[6]   It should be noted that, to avoid double counting, results from these
three studies cannot be simply added to the calculation of the climate
impacts of milk occurring in other parts of the dairy chain. All three,
for instance, include the environmental impacts from distribution and
from the activities of the dairy.

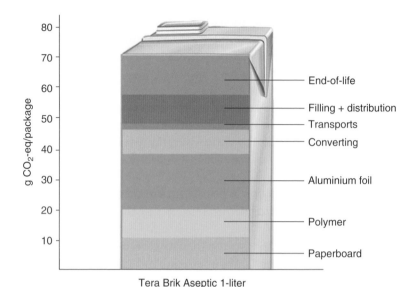

Tera Brik Aseptic 1-liter

**Figure 6.8   Indicative CO$_2$ profile of an aseptic carton.** *Source*: Tetra Pak (2007) www.tetrapak.com/environment/climate_change/co2foot print/carton_footprint/co2profile_asepticcarton/Pages/default.aspx.

over two-thirds of the weight, generates only one-quarter of the CO$_2$-eq impacts. The figure also shows that, using average European figures for recycling and recovery in 2007, roughly 20% of overall CO$_2$-eq emissions generated occur at the end of life (Tetra Pak, 2007[7]).

Reduction in the climate impacts of packaging can contribute to reductions in GHG emissions for the dairy chain as a whole, particularly in countries with an advanced distribution and retail sector. In emerging markets, meanwhile, the expansion of milk processing and the use of packaging contributes much more significantly to safe and efficient food distribution and is therefore an important support to the economic and social pillars of sustainability.

The dairy sector's drive towards sustainability will continue to be underpinned by packaging's potential to further reduce its environmental impacts and therefore those of the whole chain. While it is true that farming systems represent a much larger share of GHG emitted (US Dairy, 2008) along the chain, reduction efforts here may in many circumstances require quite some time for effective implementation. The scope is for more rapid

---

[7]   www.tetrapak.com/environment/climate_change/co2footprint/ carton_footprint/co2profile_asepticcarton/Pages/default.aspx.

environmental progress, albeit on a smaller basis, by a focus on the impact reduction potential of packaging.

The key drivers of environmental performance for optimised packaging systems (Figure 6.2) are:

- the type of packaging material chosen;

- the type of energy used;

- the choice of end-of-life (waste) treatment.


## 6.4   Global alignment of packaging requirements: implications for dairy

### 6.4.1   Business and retailers agree on packaging sustainability indicators

As a response to intensifying stakeholder and market expectations on packaging sustainability, leading global retailers together with manufacturers of consumer goods and packaging have developed a common set of definitions, indicators and metrics on packaging sustainability under the umbrella of the Consumer Goods Forum (CGF).[8] The CGF brings together over 650 retailers, manufacturers and service providers from across 70 countries with a combined sales of €2.1 trillion.

The global protocol (Consumer Goods Forum, n.d.) provides a set of indicators and metrics to offer a common language on how to measure packaging sustainability and how to communicate it along the supply chain. Companies such as Walmart, Tesco and Carrefour have played a key role in their preparation. The protocol covers environmental, economic[9] and social[10]

---

[8]   www.theconsumergoodsforum.com.

[9]   Economic indicators (e.g. total cost of packaging) are not further considered in this chapter as they are already used in business relations between producer and their packaging suppliers.

[10]   Social indicators are not further considered in this chapter because they are either covered in packaging requirements (e.g. on food safety, shelf-life) by packaging purchasers or because they are not specific to packaging only (e.g. child labour, discrimination, workplace practices) and thus are covered in wider social compliance programmes with respective certification schemes increasingly used in food supply chains (e.g. SEDEX).

**Table 6.2    Packaging sustainability indicators and metrics.**

**Environmental**

*Attributes*

| | | | |
|---|---|---|---|
| Packaging weight and optimisation | Packaging to product weight ratio | Material waste | Recycled content |
| Renewable content | Chain of custody | Assessment and minimisation of substances hazardous to the environment | Production sites located in areas with conditions of water stress or scarcity |
| Packaging reuse rate | Packaging recovery rate | Cube utilisation | Environmental management systems |
| Energy audits | | | |

*Life-cycle indicators and impacts*

| | | | |
|---|---|---|---|
| Cumulative energy demand | Freshwater consumption | Land use | Global warming potential (GWP) |
| Photochemical ozone creation potential (POCP) | Ozone depletion | Acidification potential | Toxicity (cancer) |
| Toxicity (non-cancer) | Aquatic eutrophication | Freshwater ecotoxicity potential | Particulate respiratory effects |
| Non-renewable resource depletion | Ionising radiation (human) | | |

**Economic**

| | |
|---|---|
| Total cost of packaging | Packaged product wastage |

**Social**

| | | | |
|---|---|---|---|
| Community investment | Packaged product shelf-life | Freedom of association and/ or collective bargaining | Child labour |

**Table 6.2**   (*continued*)

| Occupational health | Excessive working hours | Discrimination | Responsible workplace practices |
|---|---|---|---|
| Safety performance standards | Forced or compulsory labour | Remuneration | |

*Source*: Consumer Goods Forum (2011) Global Protocol on Packaging Sustainability 2.0: a global project by The Consumer Goods Forum. Reproduced with permission of The Consumer Goods Forum.

indicators applied over the entire packaging life cycle.[11] Built on the 'the intersection between the role of packaging and the principles of sustainability', they require packaging to:

- be able to meet market criteria for performance and cost;
- be designed holistically with the product in order to optimise overall environmental performance;
- be made from responsibly sourced materials;
- be manufactured using clean production technologies;
- be efficiently recoverable after use;
- be sourced, manufactured, transported and recycled using renewable energy;
- meet consumer choice and expectations;
- be beneficial, safe and healthy for individuals and communities throughout its life cycle;
- be able to meet market criteria for performance and cost. (Consumer Goods Forum, 2010: 7)

A list of indicators is given in Table 6.2. The CGF protocol provides detailed guidance for each indicator (e.g. definition, metric, reference standards or formulas). Examples for two

---

[11]   Packaging life cycle covers the consecutive and interlinked stages of a packaging system, from raw material acquisition or generation from natural resources to final disposal.

**Box 6.1   Environment Indicator: Packaging recovery rate**

**Definition**

The mass fraction or absolute mass of packaging recovered from all sources (commercial and residential) based on relevant waste management statistics.

1.  **Demonstration of:**

    - **Recoverability:** EN 13427 + ISO/CD 18601
    - **Material recycling:** EN 13430 + ISO/CD18604 + ISO/TR 16218 chemical recovery
    - **Energy recovery:** EN 13431 + ISO/CD 18605
    - **Composting / organic recovery:** EN 13432 + relevant ISO/CD 18606*

2.  **Recovery rate:** expressed as % of total packaging weight [% wt.] put on the market or as mass expressed by rate x total packaging weight put on the market.

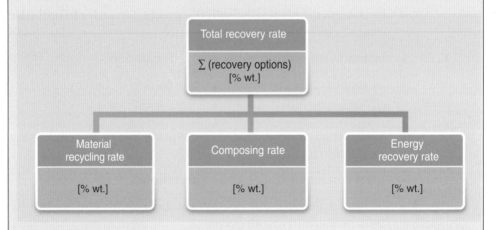

*Metric*

1.  **Recoverable** - Yes, meeting criteria or No.

2.  **Recovery rate** [% wt.] with respect to total weight of packaging placed on the market per recovery option. Total recovery rate is the sum of individual recovery rates.

*Example*

1.  Yes or No

2.  Recovery rate [% wt.]

*What to measure*
1. Determine if packaging conforms to the criteria for recoverability as per the relevant standards above. Include disclosure of material aspects of the package that would preclude recovery, for example, colour, material combinations or coatings.

2. If criteria are fulfilled, express total recovery rate as % of total packaging weight put on the market that is effectively recovered and provide the break-down per practiced recovery option.

   - **Material recycling:** measure each type of packaging produced and/or used for which national waste management recycling rates exist. Note that depending upon the packaging (type, shape, size, colour) true recycling rates might not coincide with national recycling rates for specific material or packaging category.

   - **Composting:** measure each type of packaging produced and/or used for which national waste management industrial composting rates exist. Note that in many regions the rate of composted organic waste may not coincide with the rate of composted packaging waste due to lack of acceptance.

   - **Energy recovery:** if packaging is deemed to have energy recovery value and appropriate infrastructure exists, use national waste management statistics. If data is available, measure by material type.

*What not to measure*
Packaging going to final disposal and non-recovered littering is implicitly calculated from the recovery rate and does not need to be measured separately.

*Composting and biodegradation, ASTM D6400 - 04, ASTM D6868 - 03, ISO 14855-1 or other pertinent standards.
*Source:* Consumer Goods Forum, 2011, *Global Protocol on Packaging Sustainability 2.0*: a global project by The Consumer Goods Forum. Reproduced with permission of The Consumer Goods Forum.

indicators, packaging recovery rate and global warming potential (GWP), are given in Boxes 6.1 and 6.2, respectively.

Participants in supply chains may choose from the final list the indicators and metrics they wish to use. These metrics are intended to be used for business-to-business communications on packaging sustainability, and have been tested in 22 pilot projects around the world (Raja & Hagedorn, 2010).

**Box 6.2   Environment indicator: global warming potential**

**Definition**
Global warming potential is a measure for a process' contribution to climate change. The ability of chemicals to retain heat on the earth (radiative forcing) is combined with the expected lifetime of these chemicals in the atmosphere and expressed in $CO_2$ equivalents.

**Metric**
Mass of $CO_2$ equivalents (e.g. kg $CO_2$-eq / functional unit), using the characterisation factors of the 4th assessment report of the Intergovernmental Panel on Climate Change (IPCC). A 100 year time perspective is recommended. The time perspective chosen should always be communicated together with the metric.

**Whom/What at the end am I damaging?**
Global warming will result in a net global increase of temperatures, which will be translated into very different and hardly predictable changes in climate on a local scale. These include increased or decreased precipitation, more extreme climatic events (storms, draughts), and even possibly global changes in ocean currents (Gulf Stream). This has dramatic effects on nature (modifying entire ecosystems), humans (more natural disasters, more heat-related disease, such as heart attacks, wider spread of diseases currently limited to tropical regions, such as malaria), and the economy (more natural disasters, better or worse agricultural yields, depending on the local climate).

**How do I damage?**
Emissions of greenhouse gases change the radiation equilibrium of the earth, retaining a larger amount of infrared radiation that previously was released into space. The most important greenhouse gases are water vapour and carbon dioxide ($CO_2$), which is released from combustion processes. Other potent greenhouse gases are methane ($CH_4$, from livestock farming, rice cultivation, and landfills), and nitrous oxide ($N_2O$, mainly from fertiliser application in agriculture).

**Why does it matter?**
Climate change is a serious environmental threat, with potentially dramatic impacts. A reduction of greenhouse gases is very urgent, since non-reversible change to the global climate may occur if the current amount of greenhouse gases will be emitted for only a few more years.

**What do I have to check, take into account in my supply chain?**
Impacts on global warming occur in particular if energy from fossil fuels is consumed, or agricultural activities with fertiliser use are within the system boundaries. If biogenic resources are employed, significant uptake of $CO_2$ may occur, which in LCA is accounted for as a negative emission of greenhouse gas.

**When do I have to use/select/consider this indicator?**
Global warming potential is influenced by the use of fossil resources and can be a valuable indicator to detect differences in intensity of fossil resource use or when comparing systems based on fossil resources with systems based on renewable resources.

**How specific can I interpret the resulting indicator?**
Know-how on climate change has increased drastically in the past, and the global warming potential is today a relatively reliable indicator today. Soil emissions of greenhouse gases from agriculture (changes of carbon content in soil due to cultivation practices or emissions of $N_2O$ after fertiliser application) are strongly dependent on local soil conditions, and therefore, have high uncertainties in inventory databases. Although the 100 year perspective is considered in most policy initiatives today, some consider the 500 year perspective to be more scientifically robust. Examining the 500 year perspective as a sensitivity check might therefore prove useful.

**How can I reduce uncertainty & evaluate the significance of an impact?**
Make sure agricultural processes are correctly parameterised in your inventory database.

> **Who to ask, where to look at?**
> Global warming potentials of greenhouse gases are given in the fourth IPCC assessment report (2007) and readily available in many impact assessment methods. Further guidance on carbon footprinting is provided by the World Resources Institute / World Business Council for Sustainable Development Green House Gas Initiative http://www.ghg-protocol.org/, PAS 2050 (BSI), and ISO 14067 (when available).
>
> *Source*: Consumer Goods Forum (2011) *Global Protocol on Packaging Sustainability 2.0*: a global project by The Consumer Goods Forum. Reproduced with permission of The Consumer Goods Forum.

### 6.4.2    *Implications of global packaging metrics for the dairy sector*

The dairy industry is seeking to offer a systematic response to the build-up of challenges created by expanding public and stakeholder pressures. The Global Dairy Agenda for Action on Climate Change, launched in 2009, is at the heart of this response with its commitments to use resource-efficient and lowest impact packaging. It calls for specific actions to 'increase the recycling of packaging', for the 'use of packaging with the lowest environmental impact' and to 'increase recovery of waste'.

The retailer-driven global packaging metrics, discussed above, potentially take the dairy sector's response one step further by offering a specific toolkit to measure sustainability performance in a globally harmonised way. Retailer and manufacturer members of the CGF commit to engage with their commercial partners – and these include dairies – to promote a framework for environmental improvements. If used as expected by leading retailers and food and beverage companies, the metrics will become a de facto global standard which therefore will probably apply to dairy products.

### 6.4.3    *Measuring progress towards the Global Dairy Agenda for Action*

The CGF suggests that different product groups be considered when defining a 'basket of commonly used metrics' (Consumer

Goods Forum, 2010: 10). Referring to the Global Dairy Agenda for Action, the environment criteria in the basket could be the following:

- indicators related to corporate and/or sector goals;

- indicators related to the most frequent legislative, retailer and consumer demands;

- indicators that best represent the packaging life cycle performance (lead indicators) of areas that are relevant (e.g. climate impact);

- indicators where data are generally available and relevant for stakeholders, such as recycling rate, recycled material content and renewable material content;

- indicators where sector-specific rules and guidelines on how to apply international standards are available (e.g. IDF, 2010).

Matching these criteria with the Global Dairy Agenda for Action, the drivers for packaging environmental impact reduction (see section 6.3.3) and the CGF global packaging metrics (see section 6.4.1) provide a simple formula for driving environmental sustainability for dairy packaging systems as outlined in Figure 6.9. This formula also represents the basis for selecting the basket of proposed sustainability indicators and metrics for dairy packaging systems outlined in Table 6.3.

### 6.4.4  Benefits to the dairy industry of adopting global packaging metrics

A dairy packaging assessment tool based on dairy packaging indicators and metrics used globally by leading retailers and food producers would create favourable conditions for dairy companies to measure progress against their global sector commitment and company environmental targets. Even more important perhaps, it would allow them to communicate their packaging impact reductions with major retailers easily and credibly. The benefits are therefore several:

- A harmonised approach of this type would enable the dairy industry to provide packaging information in line with retailer scorecards.

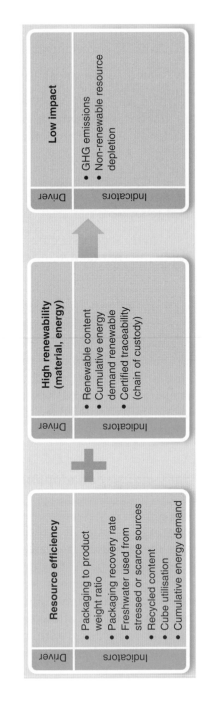

**Figure 6.9    Drivers and related key environment indicators.** *Source*: Tetra Pak (2011).

Table 6.3  Basket of proposed sustainability indicators and metrics for dairy packaging systems.

| Packaging indicator | Metric | Legal or market requirements | Related to sector goals[1] | Data generally available |
|---|---|---|---|---|
| **Attributes** | | | | |
| Packaging to product weight ratio[2] | kg packaging/kg product | | Yes | Yes |
| Packaging recovery rate | % recovered by weight of total packaging used | Yes | Yes | Yes |
| Renewable content | % of total material used/packaging component | | Yes | Yes |
| Cumulative energy demand non-renewable+renewable (CED$^{NR}$ CED$^R$) | MJ/functional unit | | Yes | Yes |
| Recycled content | % recycled content/packaging component | | Yes | Yes |
| Cube utilisation | Total volume product/transport volume | | Yes | Yes |
| Certified traceability[3] (chain of custody) | Unknown, known or sourced-certified | | Yes | Yes |
| Production sites located in areas with conditions of water stress or scarcity | Number of facilities located in areas with stressed or scarce water resources | | | Yes |
| **Life cycle impacts** | | | | |
| Global warming potential | kg CO$_2$-eq/functional unit | Yes | Yes | Yes |
| Non-renewable resource depletion | kg antimony equivalents/kg product | | Yes | Yes |

[1] Global Dairy Agenda for Action on Climate Change.
[2] Includes primary, secondary and transport packaging, including shrink wraps & slip sheets, but not pallets.
[3] Increasingly requested by stakeholders as key element of resources efficiency.
Source: Tetra Pak (2011) based on *Global Protocol on Packaging Sustainability 2.0*: a global project by The Consumer Goods Forum.

- It would reduce the response time needed to meet retailer requirements, and would provide a fact base for improved consumer communications and thus potentially for improved consumer perceptions.

- It could offer opportunities to deliver on consumer expectations as regards packaging and help in the positive positioning of dairy products.

A fact base, grounded on a common basket of indicators and metrics using commonly available data, can contribute to better understanding of the potential for reducing the impact of post-farm-gate GHG emissions. It would also improve the decision-making process and cooperation between all partners and stakeholders in the value chain: packaging suppliers, retailers, consumers and governments.

Demonstrating industry leadership by proactively managing packaging-related issues via self-regulation would allow companies to demonstrate progress without government intervention and perhaps prevent packaging-related legislation in inappropriate areas. Informing consumers and empowering them to make informed choices and participate in packaging recycling schemes may even, in addition, prevent unwarranted and costly legislation and gain government support. Finally, a common, agreed approach on packaging indicators and metrics could help drive the development of the infrastructures needed to maximise collection and recycling of milk packaging – the key stakeholder demand on packaging.

## 6.5   A company response: the example of Tetra Pak

For companies involved in the food and drink supply chain, environmental sustainability has become a key factor in remaining competitive. Tetra Pak's ongoing response to the sustainability challenge starts by recognising that the global packaging supply chain has made impressive progress in sustainability performance in recent years, has set new agendas and objectives and, in so doing, has raised the expectations made of its participants to a new level.

Staying ahead of this curve means a systematic approach to driving environmental improvements throughout the life cycle of the products and services delivered to customers. As regards dairy, this means supporting progress towards the commitments made in the Global Dairy Agenda for Action on Climate Change (2009), particularly by contributing to the Agenda's resource efficiency actions to 'increase recycling of packaging' and the 'use of packaging with the lowest environmental impact', and also to 'optimized product distribution'.[12] In its efforts in these areas, Tetra Pak uses a two-step process as the starting point: first, assembling the relevant scientific fact base, and second, understanding and assessing stakeholder demands and market expectations.

### 6.5.1   Reducing environmental impact over the life cycle

Tetra Pak applies a life cycle approach (Figure 6.10). The life cycle assessment of environmental impacts is one of the company's main tools in developing benchmarks for product development and for internal goal setting. Based on LCAs, GHG emissions are used as a lead indicator because climate change has come to be generally acknowledged as one of the most urgent global challenges. Specifically, the company focuses on three main action

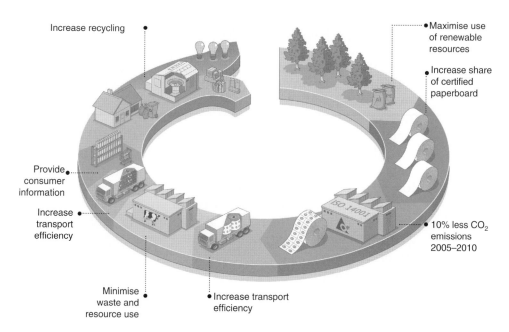

**Figure 6.10   Greening packaging supply chains: example Tetra Pak.**
*Source*: Tetra Pak (2011).

---

[12]   Global Dairy Agenda for Action on Climate Change (2009), section on 'specific actions' referring to 'resource efficiency'.

areas to reduce GHG emissions: optimising packaging material, capping GHG emissions, and sustainable sourcing.

### Optimise material used in packaging systems while ensuring they continue to meet market requirements

Tetra Pak seeks to do more with less. The main tools here are 'design-for-the-environment' standards integrated in the company's innovations and product development process. All its packaging for worldwide markets complies with the relevant European packaging environmental standards.[13] The pressure for optimisation of material use will increase in line with the growing cost of raw materials and energy supplies.

### Take specific measures to cap GHG emissions at 2010 levels

In today's market, a market leader cannot afford to be 'average' on the climate impact of its products: there is a clear incentive to outperform. Two main measures are used to lower the climate impact of packaging solutions: increasing both their renewable material content and their renewable energy content (Figure 6.11 illustrates the average material composition of a Tetra Pak carton).

These measures underpin Tetra Pak's 2020 climate goal: the objective is to cap the climate impact along its value chain at 2010 levels in spite of growth in output. Already cartons have among the lowest climate impact for packaging used for milk. Tetra Pak has made a commitment to reduce this further by increasing the share of renewable material used in the manufacture of cartons. It is exploring the potential for material innovation and specifically how bio-based polymers can substitute for the fossil-based polymers used at present in its milk cartons. In 2011 the company began purchasing from Braskem, one of America's largest chemicals and plastics suppliers, 'green' or biomass-based high-density polyethylene (HDPE) for package closures. The ethylene converted into 'green' polyethylene is

---

[13]   EN 13428: 2004, Packaging – Requirements specific to manufacturing and composition – Prevention by source reduction; EN 13430: 2004, Packaging – Requirements for packaging recoverable by material recycling; EN 13431: 2004, Packaging – Requirements for packaging recoverable in the form of energy recovery, including specification of minimum inferior calorific value; see also www.cen.eu/cen/Sectors/Sectors/TransportAndPackaging/Packaging/Pages/PPW.aspx.

**Figure 6.11    Typical material composition of Tetra Pak packages.**
*Source*: Tetra Pak (2010).

made from sugar cane sourced and produced in Brazil.[14] Caps made of renewable plastic from sugar cane were first used by Nestlé in Brazil. This is the beginning of a new phase of exploring further increases in the renewable materials' share of Tetra Pak cartons, which aim at delivering highest food protection at lowest environmental impact. Substitution of aluminium is another area of exploration: aluminium, while representing just 4% material weight of an aseptic carton, contributes well over 40% to the packaging material's climate impacts.

As regards manufacturing, the company's plants have achieved an absolute 10% reduction in $CO_2$ between 2005 and 2010 (in addition to years of energy efficiency programmes prior to 2005). This represents a decoupling of $CO_2$ emissions from growth in production output. It was achieved through energy savings in the factories as well as through the use of renewable energy and corresponds to more than a 30% relative reduction compared to 'business-as-usual' since 2005. These reduction achievements have received recognition through Tetra Pak's partnership with the WWF in its Climate Savers

---

[14]  www.plasticstoday.com/articles/bioplastics-tetra-pak-braskem-bio-based-hdpe-1130.

programme.[15] The company is committed to cap climate impact from its operations and products, the packaging and equipment it supplies, by 2020 at 2010 levels.

The 2020 goal is a new departure, covering the company's entire value chain. Achieving this will require contributions to environmental performance from its suppliers, in addition to continuously improving the energy consumption of its own operations. Already about 70% of the energy used in the European paper mills producing paperboard for the beverage carton comes from renewable energy (Alliance for Beverage Cartons and the Environment, 2005). In other parts of the world, there will be reliance on energy efficiency increases and the availability of cleaner and renewable energy in the locations where its suppliers produce. GHG emissions will be a stronger factor in Tetra Pak's purchasing requirements. An overall value chain target for reduction in environmental impact also requires success in driving its recycling efforts. The greater the number of used cartons recycled, the fewer go to waste disposal in landfills where GHG emissions are generated.

### Drive sustainability through the supply chain: sustainable sourcing

Tetra Pak increasingly demands of its suppliers that they source from forests with certified management systems. This is based on the recognition that managing forests responsibly enhances their natural carbon sink benefit. In 2010, 40% of the paperboard the company purchased was certified by the Forest Stewardship Council[TM][16] and more than 18 billion Tetra Pak cartons delivered in 2011 carried the FSC label (Figure 6.12). Since 2010 Tetra Pak customers have launched FSC-certified packaging in many markets – including Germany, France, Spain, Sweden, Denmark, China, Thailand, Switzerland, Belgium and the Netherlands. The progress made in 2011 is a further step towards the final goal of having 100% FSC-certified paperboard supply.

---

[15]  wwf.panda.org/what_we_do/how_we_work/businesses/climate/climate_savers/partner_companies/.

[16]  The Forest Stewardship Council (FSC) is an organisation that has created consensus standards with environment groups, governments and industry. It has developed a set of very exacting standards. The FSC label is increasingly recognised globally as the premium standard in forest management; see also www.fsc.org.

**Figure 6.12    FSC-labelled Tetra Pak milk carton.**
*Source*: Tetra Pak (2010).

The world's leading beverage carton producers, including Tetra Pak, have made a sector-wide commitment, recognised by the WWF and the EU Commission, to achieve 100% traceability by 2015 of the wood used for the paperboard in their carton packages. Tracing wood back to its forest of origin is the first step in the long road towards responsible management of forests which, it is recognised, play a key role in the mitigation of climate change. By the end of 2011, over 74% worldwide of Tetra Pak's packaging material converting plants were FSC chain-of-custody certified.

In addition, Tetra Pak is increasingly pressing suppliers of the non-paperboard materials used in its cartons to demonstrate responsible sourcing practices. Positive progress has been made in this direction for bio-based polymers; however, for conventional non-renewable materials no certification systems are yet available.

### 6.5.2  *Increasing recycling*

Consumers see recycling as the action with the greatest positive impact on the environment in Europe and globally. From a dairy product perspective, this means maximising the recycling of packaging for milk and liquid dairy products.

The company is driving recycling in all markets and in all geographies. In the European Union an average recycling rate of 37% (Alliance for Beverage Cartons and the Environment, 2012) and a total recovery rate of 68% (energy recovery as well as recycling) was achieved in 2011. Globally about 22% of all Tetra Pak packages put on the market were recycled (Tetra Pak, internal statistics).

## 2020 recycling goal

The company's goal is to double its current global recycling rate by 2020. It works in partnerships with customers, retailers, governments and local authorities, waste collectors, recyclers and NGOs to support the development of collection and recycling infrastructures and to encourage consumers to participate in recycling programmes.

Milk cartons collected from households are mostly recycled in paper mills. The paper recycling process separates the paperboard from the carton's polymer and aluminium layers, enabling the high fibre quality in the paperboard to be used for new products such as secondary packaging for food and consumer goods, tissue paper and office stationery.

The separated non-fibre components are typically used in industrial processes where aluminium is recycled for use in a wide range of products and the polymers are recovered as energy. Applications for the non-fibre components differ and depend on local conditions regarding volumes, technologies available, and prices of secondary raw materials.[17]

It is company policy to provide support to recycling technology developments which are adapted to local conditions and that create sustainable markets for both the used paperboard fibres and the polymer and aluminium components recovered. Keeping the maximum of material within the economy and maintaining the value, quality and quantities of the material collected are key objectives; both depend largely on consumer participation and the infrastructures available.

The emphasis is on securing recycling capacities, on finding applications for the different material components of the packages, and on supporting packaging waste collection. An additional

---

[17]    Tetra Pak, http://campaign.tetrapak.com/lifeofapackage/recycling.

challenge in developing markets is the incorporation of largely informal waste collection, sorting and trading sectors into the mainstream economy and improving their operating conditions. In these countries, Tetra Pak collaborates with government, NGOs and customers, and shares the knowledge gathered on these developments around the world both internally and with its partners. In each of Tetra Pak's eleven market clusters worldwide there are specialised experts in the setting up of collection and recycling infrastructures.

### 6.5.3  *Engaging with stakeholders*

In the newly emerging sustainability landscape – where many markets tend to be regulated by an intertwined mixture of public policy, consumer expectations and resultant stakeholder requirements – a sine qua non for credible environmental performance is engagement with all key stakeholders. Only with such support and involvement can the targets the company seeks to achieve over its business life cycle, and in other value chains in which it participates, be attained.

A priority focus for Tetra Pak is engagement with its supply chain. It works with its paperboard suppliers to ensure the wood fibre used in its products is sourced responsibly. The company partners with WWF on a sustainability agenda for forestry, and since 2006 has been part of the WWF Climate Savers programme. It works systematically with customers, public authorities, local NGOs and industry coalitions to drive collection schemes needed to increase recycling rates.

An active participant in the dairy sector's supply chain, Tetra Pak supports the Global Agenda for Action on Climate Change (see section 6.4). It partners with retailers, e.g. with Carrefour on a pilot project on UHT milk packaging in France which tested the sustainability packaging metrics developed by the Consumer Goods Forum (see section 6.3.1), another significant stakeholder.

The greening of packaging supply chains is seen by the company as a key ingredient in efforts to create more sustainable consumption and one which will contribute to the overall environmental performance of the dairy sector. Environment will become an ever more important factor in successfully competing in the marketplace. Tetra Pak is committed to maintaining measurable improvements that further underpin the sustainability goals of the dairy chain.

## 6.6   Outlook: growing importance of economic and social pillars of sustainability

The focus on the role of packaging in contributing to the sustainability of the food and drink sector is sharp and likely to intensify. As regards dairy, this is explained by factors specific to the sector, outlined above, and also by more general trends. The structural growth in demand for food worldwide is an important development, in turn leading to the production and distribution of more dairy products, more packaging and more used packaging to be sustainably managed. Persistent consumer expectations on recycling, and the stepped-up screening by retailers of their supply chains, are significant.

The importance attached to the role of packaging by major stakeholders is amplifying the demand for improved environmental standards for the dairy value chain. Retailers in particular are now seen as gatekeepers to the sustainable marketplace and increasingly as quasi-regulatory guarantors of sustainable consumer choice. Reducing climate impact is widely accepted as an important touchstone of sustainability performance.

Adding to these pressures, a new sustainability priority is emerging worldwide. Food security and surging food prices, notably for food staples, are creating a new focus on food losses as a priority area for action in responding to a broader sustainability imperative. This asks the question: can the planet be considered sustainable if its capacity to feed a growing population is diminishing?

The response to this question involves fresh challenges both for dairy and for packaging as an enabler in preventing food losses. An indication of the scale of the problem in developed markets is given in an EU study which suggests an annual total of 89 million tonnes of food loss generating 170 million tonnes of GHG impacts, much of it created by households and most of that avoidable (European Commission, 2010).

Research on the broader international picture published by the FAO examined how packaging, along with other factors, might help solve the problem, particularly in developing countries. 'Roughly one-third of food produced for human consumption is lost or wasted globally, which amounts to about 1.3 billion tonnes per year' (FAO, 2011[18]). For meat and dairy the yearly global quantitative food losses and waste are estimated at

---

[18]   www.fao.org/docrep/014/mb060e/mb060e00.pdf.

roughly 20% (FAO, 2012[19]). Particularly for developing countries the role of appropriate food packaging solutions are seen as
key. 'Losses at almost every stage of the food chain may be
reduced by using appropriate packaging. The global food packaging industry has a lot to contribute not only in addressing
food losses but also in ensuring food safety as well as enhancing
global food trade, which is a key to economic development of
varying economies (FAO, 2011[17]). The sustainability challenges
confronting the dairy and packaging sectors in the years to come
will become more complex. Today's agenda is one with a strong
environmental emphasis. This pressure will be maintained but,
in a global economy marked by food security concerns and population growth, it will merge into a definition of sustainability
whose economic and social dimensions are likely to become
more prominent and create new demands.

---

[19]   www.fao.org/docrep/015/i2776e/i2776e00.pdf.

# References

Alliance for Beverage Cartons and the Environment (2005) *Working with Nature*. Brussels: ACE.

Alliance for Beverage Cartons and the Environment (2012) Beverage carton recycling rates 2011. Press release 2 July 2012; www.beverage carton.eu.

Brundtland Commission (1987) *Our Common Future: Report of the World Commission on Environment and Development.* Published as Annex to United Nations General Assembly document A/42/427, Development and International Co-operation: Environment. 2 August 1987.

Consumer Goods Forum (n.d.) *Global Protocol on Packaging Sustainability 2.0: A global project by The Consumer Goods Forum,* http://globalpackaging.mycgforum.com/allfiles/GPPS_2.pdf.

Consumer Goods Forum (2010) *A Global Language for Packaging and Sustainability: A framework and a measurement system for our industry.* June 2010.

Council of the European Union (2006) *Review of the EU Sustainable Development Strategy (EU SDS) 10117/06,* register.consilium.europa. eu/pdf/en/06/st10/st10117.en06.pdf.

European Commission (2010) *Preparatory Study on Food Waste Across EU 27.* European Commission, DG Environment, October 2010.

European Environmental Bureau (2009) *The blueprint for sustainable consumption and production.* Brussels: EEB.

Food and Agriculture Organization of the United Nations (2010) *FAO Report: Greenhouse gas emissions from the dairy sector: a life cycle assessment.* FAO, Animal Production and Health Division. www.fao.org/docrep/012/k7930e/k7930e00.pdf.

German Federal Ministry for the Environment, Nature Conservation and Nuclear Safety (1991) *German Waste Act, Ordinance on the Avoidance of Packaging Waste. (Verpackungsverordnung).* June 1991.

German Federal Ministry for the Environment, Nature Conservation and Nuclear Safety (2005) Last amendment, German Waste Act, Ordinance on the Avoidance and Recovery of Packaging Wastes (Packaging Ordinance – Verpackungsverordnung). *Federal Law Gazette,* I (27 May 2005): 1407.

GfK Roper Consulting (2009) *GfK Roper Reports Worldwide: Mood of the World 2009.*

Global Dairy Agenda for Action on Climate Change (2009) www.dairy sustainabilityinitiative.org/Files/media/Declarations/Declaration_ Final-Text_English-18-June-2009.pdf.

GlobeScan Incorporated (2009) Environmental Opinion Research among key stakeholders, decision makers and influencers in key market globally, commissioned by Tetra Pak.

Government of France (2010) *Journal Officiel De La République Française,* 13 juillet 2010, LOI no 2010-788 du 12 juillet 2010 portant engagement national pour l'environnement, articles 199 and 201.

International Dairy Federation (2010) A common carbon footprint approach for dairy: the IDF guide to standard lifecycle assessment methodology for the dairy sector. *Bulletin of the International Dairy Federation*, no. 445. Brussels: IDF. www.IDF-LCA-guide.org.

NABU (Naturschutzbund Deutschland e.V.) (2010) NABU vergibt 'Grünen Einkaufskorb' Gut durchdachte Konzepte für ökologischen Einkauf, www.nabu.de/themen/nachhaltigkeit/ressourcen/gruenereinkaufskorb/13150.html. 30 November 2010,

People's Republic of China (2009) *Recycling Economy Promotion Law of the People's Republic of China*, effective 1 January 2009, www.lawinfochina.com/Law/Display.asp?Id=7025.

Petrie, Edward M. (2011) *The Future of Sustainable Packaging to 2020: Convenience vs the Environment*. PIRA International. January 2011.

Raja, Sonia, Hagedorn, Rudy (2010) *Global Packaging Project: Pilot Findings*, Paris. 14 October 2010.

Sevenster, Maartje, de Jong, Femke (2009) *A sustainable dairy sector: global, regional and life cycle facts and figures on greenhouse-gas emissions*. CE Delft.

Tetra Pak (2009) *Tetra Pak Dairy Index*, Issues 1 and 2, www.tetrapak.com/food_categories/dairy/index/Pages/default.aspx and www.tetrapak.com/food_categories/dairy/index/whatisdairyindex/pages/default.aspx.

US Dairy (2008) U.S. Dairy Sustainability Initiative, *A roadmap to reduce greenhouse gas emissions and increase business value*. Innovation Center for US Dairy, December 2008.

van Otterdijk, Robert (FAO officer for the SAVE FOOD global initiative), www.ceepackaging.com/2011/01/'31/food-waste-to-be-examined-at-interpack.

von Falkenstein, Eva, Wellenreuther, Frank, Detzel, Andreas (2010) *LCA studies comparing beverage cartons and alternative packaging: can overall conclusions be drawn?* Springer Verlag. 27 July 2010, http://springerlink.com/content/d1468l2x4r317506/fulltext.pdf.

World Business Council for Sustainable Development (2010) *Vision 2050: The new agenda for business*. Geneva: WBCSD. Summary at www.wbcsd.org/web//projects/BZrole/Vision2050_Summary_Final.pdf.

Young, S., Ciummo, V. (2009) *Shopper Sentiments and the Environment: A Cross-Cultural Comparison, Part 4 Shopper Experience in Retail Shopper*, Esomar.

# 7

# The business case for sustainable dairy products

Jaap Petraeus

Royal FrieslandCampina, Amersfoort, The Netherlands

**Abstract:** A business model on how to approach sustainable dairy products is presented. Elements such as pricing power and cost savings, and employee productivity, brand reputation and new markets, are discussed. The chapter shows how different food companies are expected to cope with future demands for sustainability.

**Keywords:** branding, business case, marketing, shareholders, society, stakeholders, sustainable targets

## 7.1 Introduction

The previous chapters contained information on how to make the production process (dairy processing) more sustainable, for example through concentration by evaporation, drying, cleaning and cooling. The aim is to allocate resources, such as energy and water, carefully and cost-efficiently, and to minimise waste creation. The measures in question are being taken at production level. Often they are technical or organisational, and are designed

*Sustainable Dairy Production*, First Edition. Edited by Peter de Jong.
© 2013 John Wiley & Sons, Ltd. Published 2013 by John Wiley & Sons, Ltd.

to lead to more efficient production. Norms and guidelines are based on legislation and regulations, such as the European IPPC (Integrated Pollution, Prevention and Control) Directive and the accompanying BAT/BREF (Best Available Techniques and Reference Document) Directive. The role of government is particularly important in relation to the processing component. States set out specifications governing the efficient use of raw materials through statutory regulations, and check that these requirements have been met. Environmental licences often depend on prescribed essential preconditions. The buyers and consumers of the products play almost no role at this stage, except in the case where they live close to the dairy plant.

In recent years, there has been a significant shift away from process-based policy and towards an integrated product-based policy. As a result, policy now focuses much more strongly on the end-product and far less on the separate processes to create it. This chapter discusses this shift and the effect it has had on efforts to increase the sustainability of dairy products. For many dairy businesses, the end-products are the tangible result of activities carried out by all links in the chain. They can be used to show that production is sustainable. Society and consumers/buyers are the main stakeholders in the focus on sustainable products. They impose requirements governing the quality of the product and are partly responsible for ensuring that the sustainability process satisfies the requirements that consumers expect of the producers. The consumer/buyer also determines the added value of the sustainability process relating to the product. This chapter specifically discusses the compilation of the business case for making products more sustainable and the role of various stakeholders in this process.

## 7.2 From a process-driven to a product-driven approach

### 7.2.1 Process-driven approach

The manufacture of (dairy) products takes place in a chain. Figure 7.1 shows the dairy production chain in simplified form.

The feed industry supplies feed for the dairy farmer's herd. The cow converts grass, maize and feed into raw milk. The dairy farmer delivers the raw milk to the dairy processing industry and the processing industry manufactures dairy products, which

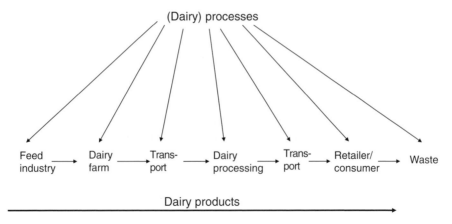

**Figure 7.1    Links in the dairy chain.**

are sold to the consumer through shops and supermarkets. The residual products are processed into animal feed, used as digester material to produce green energy, incinerated or disposed of in landfill. In the simplified chain, a large number of smaller links are needed, all of which can also contribute to making the product more sustainable.

Improving each of the production processes requires ongoing attention from the various links in the chain. Mostly each link works to maximise its own process efficiency. Consuming scarce raw materials efficiently and making cost savings are the main drivers. Every link in the chain is working to improve the efficiency of energy and water consumption, to reduce the production of wastewater and waste materials, and to minimise odour and noise. These are issues that have been high on the agenda of dairy companies for the last 20 years. The measures taken were motivated chiefly by a desire to reduce costs and respond to government regulations, in the form of licence requirements, for example. Coordination between the various links in the chain was minimal.

It is not easy to fully implement all these measures in the endproduct. For most companies, making production processes more sustainable is not an end in itself but the result of various considerations (statutory requirements, cost savings and so on) that must be weighed up by those who are responsible for production. If the law makes a measure compulsory, then there is no choice. If the measure has a short payback time (less than three years), a dairy company will often make a quick decision in favour. However, if the payback periods within the production environment are longer, then the decision is less clear; increasing the sustainability of the production process is often difficult to capitalise because of the long payback period and is therefore not always included in the equation.

In addition, to improving the efficiency of the process, considerable attention has been paid in recent years to making production processes more manageable. A structured approach is required. Many companies use the ISO 14001 (environmental protection) and OHSAS 18001 (occupational health and safety) systems to achieve this. Implementation of these care systems helps companies to manage the identified risks, make ongoing improvements to their processes and keep up to date with legislation, regulations and licences.

### 7.2.2  Product-driven approach

In contrast to the process-driven approach, a product-driven approach imposes requirements on the end-product, and the chain as a whole is responsible for meeting them. The separate links in the chain must work together to create a sustainable product. Demand arises from the market or is imposed from above by the government. Examples include requirements limiting the levels of $CO_2$ emitted by vehicles and the EU Directive regarding the energy consumed by electrical equipment. The Dutch government also makes specifications concerning sustainable procurement. In its policy on sustainable production and consumption, the EU imposes requirements on the products purchased by the governments of the member states in their Green Public Procurement Policy.

Marketing managers are responsible for drawing attention to ongoing changes in consumer demand and for translating these changing requirements into products.

One example which may help to clarify the difference between a process-driven and a product-driven approach is that the Dutch government prescribes that (production) processes must reduce their energy consumption by 2% each year. However, the measure taken in one process could cause a rise in energy consumption in another (production) process. The government would therefore be better to insist that the end-product (e.g. 1 kg of milk or 1 kg of cheese) should require 2% less energy to produce. All the links in the chain would then be obliged to work together to achieve this 2% reduction. This would reduce chain inefficiency.

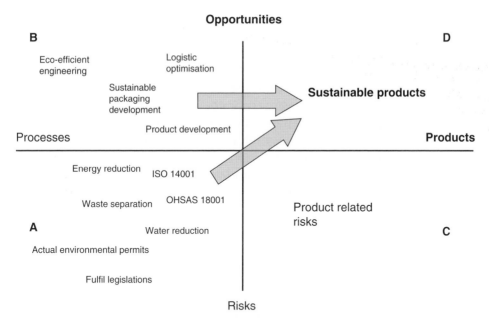

**Figure 7.2    Relationship between processes, products, opportunities and risks.**

The differences between processes and products are shown in Figure 7.2.

The horizontal axis leads from the processes to the products. The vertical axis leads from the risks to the opportunities. The left-hand side of the diagram represents operating processes. The supply chain organisation often takes the lead. It is responsible for ensuring that the operating processes are carried out (cost)-efficiently, such that the limiting conditions for sustainability are upheld. They produce what is determined by the marketing and sales department. The supply chain organisation must ensure that the process is implemented and that any risks to people and the environment are kept to a minimum. They must also comply with legislation and regulations. In order to be able to manage the processes, they use management systems such as ISO 14001 (environmental protection) and OHSAS 18001 (occupational health and safety). This approach is shown in the lower left quadrant. Communication about these issues is technical and is often not suitable for a wider public. Efforts must be made to avoid negative communication (e.g. local residents complaining about noise or odours), which often compels the government to step in. If the company works proactively with the government and upholds the agreements it has made, negative publicity can usually be avoided.

There are also many opportunities to make processes more sustainable. These opportunities are shown in the upper left quadrant. In particular, the design of new processes contains many opportunities that are often under-utilised. Together with suppliers, companies can develop new, energy- and water-efficient processes which can often generate substantial efficiency gains. There are examples of efficiency improvements of more than 20% in energy and water consumption. Sustainability benefits can also be made in the packaging and product design processes, e.g. in the method of packaging design or through optimum loading.

The right-hand side of Figure 7.2 represents the products. Here, the marketing department usually plays a decisive role. On the risk side (lower right quadrant) are the sustainability claims that cannot be realised. These include false claims relating to sustainability (so-called 'greenwashing') and blaming and shaming, e.g. by non-governmental organisations (NGOs). A well-known example was the barricading of the headquarters of Fonterra by Greenpeace in 2010 because of the company's use of palm kernel oil in cattle feed. Like the risks on the process side, these risks must be recognised early on, so that adequate measures can be taken. The damage caused by failure to do so (including to the company's reputation) is often many times greater on this side than any negative publicity on the processing side. 'Activist' NGOs use this side as a way of exerting pressure to achieve their ideological aims. Companies are often forced to meet their conditions if there is a risk that their reputation (and that of the brand) might be damaged.

The most attractive quadrant for companies in relation to sustainability is the one containing products and opportunities. In this upper right quadrant, value can be added to products through sustainability, or else completely new products can be developed which exploit the demand for more sustainable goods. This value mostly need not generate a direct income, but it can be profitable in other ways. This is discussed more fully below. Before a company can market sustainable products, however, all the other quadrants must have been addressed and must provide adequate support for the products. If this has not been adequately ensured, sustainability becomes an unmanageable risk and can only result in damage. Ongoing improvements to production processes can thus be seen as an essential prerequisite if sustainability is to be reflected in the end-product. Section 7.3 describes an approach that can be used to make processes and products more sustainable.

## 7.3 Success factors for creating more sustainable processes and products

Increasing sustainability is an open-ended process. It calls for companies to adopt a different way of thinking. It should be made part of everything the company does (and does not do) and must, as it were, become embedded in its genes. A comparison is often drawn with food quality. In the 1990s, there were regular debates about the quality of food. Food companies acknowledged the risks associated with an inability to trust the quality of the food they produced. Quality systems such as ISO 9000 were introduced as a result. Companies became well-known (and infamous in some cases) in trade journals for the quality of their food, or lack of it. The need to guarantee the quality of products is now accepted as a given. Companies can no longer gain a competitive advantage over their rivals by introducing a quality system. The concept of sustainability may undergo a similar process. By adopting a proactive approach, companies are able to distinguish themselves from their competitors.

### 7.3.1 Stages in the sustainability process

There is no general blueprint for compiling a company sustainability policy. However, experience has shown that the following stages are essential for a successful approach to such a policy:

(1) Theme analysis in markets.

(2) LCAs to identify sustainability issues.

(3) Sustainability as part of the company's vision and mission.

(4) Dialogue with stakeholders, e.g. NGOs.

(5) Scientific back-up.

(6) Transparency and communication.

# Extended theme analyses in the markets of FrieslandCampina

Globle themes in relation to the dairy chain

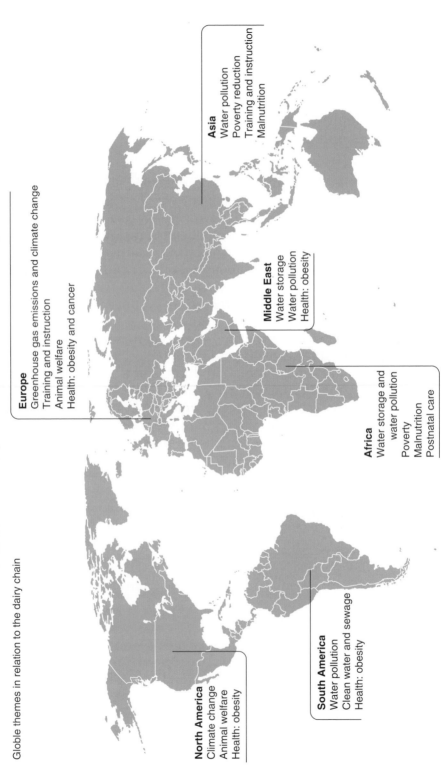

**Asia**
Water pollution
Poverty reduction
Training and instruction
Malnutrition

**Europe**
Greenhouse gas emissions and climate change
Training and instruction
Animal welfare
Health: obesity and cancer

**Middle East**
Water storage
Water pollution
Health: obesity

**Africa**
Water storage and
water pollution
Poverty
Malnutrition
Postnatal care

**North America**
Climate change
Animal welfare
Health: obesity

**South America**
Water pollution
Clean water and sewage
Health: obesity

**Figure 7.3   Example of a theme analysis.** *Source:* Royal FrieslandCampina.

## Theme analysis in markets

Sustainability is a broad concept. It is therefore important to divide it into different areas to maintain focus and enable effective agenda setting. This can only be done by conducting a detailed theme analysis in the markets in which the company operates. A theme analysis (Figure 7.3) will identify which social issues relate to sustainability in the specific market in question. These issues can be analysed on the basis of available government policy, and above all through the programmes of local NGOs. Moreover, international NGOs often cluster their policy issues by region.

## Life cycle analysis to identify sustainability issues

Life cycle analysis (LCA) is an important tool for identifying risks and possible improvements in the chain. An LCA surveys the entire chain and calculates the sustainability effects of each link. A limited LCA can be drawn up exclusively for greenhouse gases or carbon (a 'carbon footprint'), to identify the greenhouse gases that are released across the chain. An LCA can also involve more than one sustainability effect (such as water, biodiversity, atmosphere, etc.). These analyses are often more extensive and elicit a number of subjects for discussion. Analysing different magnitudes of sustainability is and will continue to be difficult. In practice, calculating a carbon footprint appears to give a good indication of the contribution that the various links make to the overall chain.

Figure 7.4 gives the example of greenhouse gas emissions involved in the production of 1 litre of milk. Transportation, packaging and retail/consumers account for only a small proportion of the whole.

## Leadership, vision and mission

Demonstrating leadership, vision and a mission in relation to sustainability are key prerequisites for the successful implementation of sustainability. If the company CEO communicates the importance of sustainability, this will have an incentivising effect within the organisation. Sustainability often does not generate short-term income, but it can inject energy and motivation. Companies with vision see sustainability as a precondition for long-term continuity. In many companies, sustainability arises from within, for example in an external communications/public

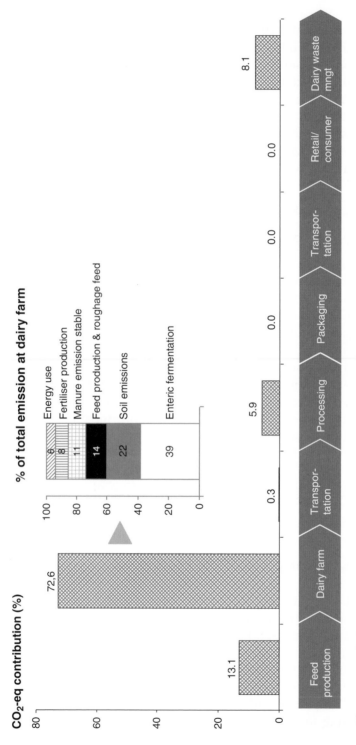

**Figure 7.4  Calculation of the carbon footprint of some links in the dairy chain.**

affairs department or a sustainability facilities department. This can work well in itself, but it takes longer to generate support within the organisation. Clear communication from the CEO (verbally and by example) appears to be an important step towards sustainability. Showing commitment to the issue and reflecting confidence in the process both encourage acceptance.

## Dialogue with stakeholders

Stakeholders are important because they provide the 'licence to produce'. Dialogue with stakeholders is an essential part of corporate social responsibility (CSR) to fit the policy to the interests of civil society. Specifically, NGOs are significant stakeholders. They represent a certain section of society and are therefore an important point of address for companies. NGOs are asked by their members to meet the goals that have been formulated (e.g. better nature conservation and human rights, opposing the felling of tropical rainforests) and they must also look for partners to help them meet these goals. Companies influence the goals of NGOs and can take measures to enable NGOs to reach these goals. Agreeing and implementing targets and measures enables shared goals to be met.

Cooperation with NGOs to attain the jointly formulated goals involves advantages for both parties. NGOs often have considerable knowledge of local situations and can readily exploit any improvements. Companies have knowledge of marketing, which they can apply to achieve the shared goal. Respect for each other's standpoint is an essential rule which all parties must respect in full. Marketers who simply want to go for the quick win (slapping the logo of an NGO on the package) may undermine a laboriously established relationship of trust in a very short space of time. Conversely, there are many examples of NGOs bringing their personal preferences to bear, which can erode public confidence in a company. Cooperation between companies and NGOs therefore calls for strict discipline on both sides, coupled with an open and transparent attitude. Many companies and NGOs still have to learn this lesson. On the other hand, cooperation provides major opportunities for both parties to reach their sustainability goals faster than they could if they were acting alone.

Box 7.1 gives an example of a partnership between companies and NGOs aimed at reducing the clearance of tropical rainforests for soy cultivation.

**Box 7.1    Practical example: Sustainable soy for dairy feed**

Soy is used as an ingredient in food (in the form of soy oil) and in animal feed (in the form of soy kernels, the residue that is left over when the oil has been pressed).

Soy cultivation in Latin America grew explosively between 2000 and 2010. The result was that large tracts of environmentally sensitive land were taken into agricultural use. This included the Amazon Basin, some of which fell victim to the high global demand for soy. Dairy and meat processing companies began to express concern about the consequences of including soy in the feed of animals used to produce meat and dairy products. NGOs were also calling these large companies to account for the unsustainable impact of soy. Large corporates such as Unilever, FrieslandCampina and Vion responded by concluding a partnership in the soy chain to promote more sustainable soy cultivation. They also set up the Round Table for Responsible Soy (RTRS) with NGOs. The retail industry later joined these initiatives and feed companies were invited to do the same. This chain-based collaboration is now gradually ensuring that more sustainable soy is included in animal feed. The Netherlands and Belgium are frontrunners, and other European countries are following suit. Asia is still lagging behind but this could change very quickly.

### Scientific back-up

Science and research have a crucial role to play with regard to sustainability. Companies need to implement measures based on scientific evidence. Initiatives in the sphere of sustainability generally have a long-term effect. This is difficult for companies, because it involves a high degree of uncertainty. The role of science is to identify these long-term sustainability effects and to give companies an idea of what is expected of them in the form of specific measures. Past examples include the problem of acidification ('acid rain') and the hole in the ozone layer. Science identifies the effects and the substances responsible for them and puts forward appropriate measures (catalytic converter, banning propellants in aerosol cans, etc.). Politicians and industry have taken measures to minimise these effects. The research community is currently focusing considerable attention on emissions of greenhouse gases and their effects on the climate. Science provides the evidence enabling the necessary measures to be taken.

### Transparency, accountability and external communication

Another prerequisite of a responsible sustainability policy is transparency and communication. The company must not be an inaccessible bulwark which decides what is best for the global community behind closed doors. It must be prepared to be accountable for its actions and enter into a public debate. Annual forms of accountability based on sustainability reports are a key instrument. Communication with stakeholder organisations is another important step towards transparency.

Based on the six steps described above, companies can develop sustainability policies in which ongoing improvements form an integral part.

## 7.4   Implementation of sustainability within the company

### 7.4.1   Role of management

Communicating the importance of sustainability is a key management role. For many companies, sustainability is a difficult concept to define. Personal motivation and communication by the CEO help to increase support for sustainability within the organisation.

Setting goals is a next step that has to start at the very highest management level and must then cascade down to the rest of the organisation. A growing number of companies are choosing to adopt quantifiable sustainability targets, which take account of the risks the company would run if it were to make insufficient progress in this area.

### 7.4.2   Generating support and building practical evidence via a stepwise approach

Specific examples of business cases in which sustainability plays a role are important for use as demonstration projects within the organisation. In many companies, there are often one or two activities that can act as a pilot. Which activities they are will depend on customer demand or the opportunities that are available to bring a particular product to the market. It is advisable to develop practical evidence for a specific product (as an innovation) and use it to gain knowledge and experience. It is better to learn with one or two products than to try to make the entire product portfolio more sustainable all at once. Attention from management is vital since different departments are often involved (e.g. production, logistics, marketing and sales). Management can formulate an objective, for example the sale of 100 tonnes of sustainably produced product. All the parties must then contribute to making the product more sustainable than that of its rivals (and marketing it as such). The marketing department must study opportunities on the market and define the conditions which the sustainable product must meet. Procurement and production must ensure that the product satisfies these criteria. The company must then team up with its customers to look for ways to position the product, after which marketing and sales take the driving seat. In practice, this is frequently a matter of trial and error, but the learning process and the demonstration effect are worth the effort.

### 7.4.3  The role of stakeholders, including NGOs

Stakeholders (Figure 7.5) can influence company policy. For most companies, the most important of these stakeholders have traditionally been the shareholders. Their interests usually focus on the company's – often short-term – economic profitability. However, sustainability cannot be directly translated into an

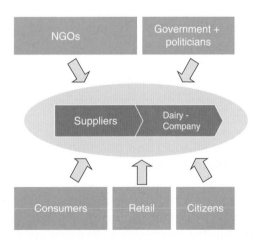

**Figure 7.5   Who are the stakeholders?**

economic return; it has more to do with responsible enterprise and hence an economic return over the longer term. Dairy cooperatives are used to working with a large group of stakeholders (dairy farmers), who are both shareholders and raw material suppliers. The aim of the cooperatives is to provide a long-term assurance that the shareholders' products will be marketed. Farming cooperatives are therefore also used to having a longer horizon than listed companies. Members of cooperatives are in many cases looking to hand on their farms to their children and grandchildren. This makes sustainability a topic that is quite easy for cooperatives to broach.

One key group of stakeholders that has become important for companies in recent years is civil society organisations or NGOs. The role of these stakeholders in particular has undergone radical changes over the past few years. In the past, companies were generally not very willing to enter into dialogue with NGOs due to the fact that they were an unknown quantity and lacked experience. For example, the decomissioned Brent Spar oil storage buoy, which Shell wished to dispose of in the deep ocean, was occupied by Greenpeace, demonstrating to Shell that, while they were technically in the right, they could not as a company compel public acceptance. Major international companies have therefore begun to adopt a much more transparent approach vis-à-vis NGOs in recent years. For their part, many NGOs have recognised that working in tandem with companies can often make their goals much easier to achieve. The knowledge and expertise embodied in companies can be used by NGOs to meet shared objectives. A good example of cooperation between NGOs and companies are the round table conferences on sustainable palm oil, sustainable soy (see Box 7.1) and sustainable sugar. All the parties involved know that not all their goals can be met, and they are therefore prepared to work towards a compromise.

It is vital for companies to gain a good idea of which stakeholders are important to them. A stakeholder analysis is a useful tool for achieving this. Such an analysis can be carried out via the AA1000 SES method (AccountAbility, 2011).

Figure 7.6 shows in three stages how stakeholder analysis works. In the initial phase, an overview is compiled of all the relevant stakeholders. A cluster is then drawn up of the nature and influence of the various stakeholders. The sustainability themes that are important to the company in the short and long term are weighed against their relevance for the stakeholders. Both analyses are then combined to create a diagram in which the activities per stakeholder are formulated.

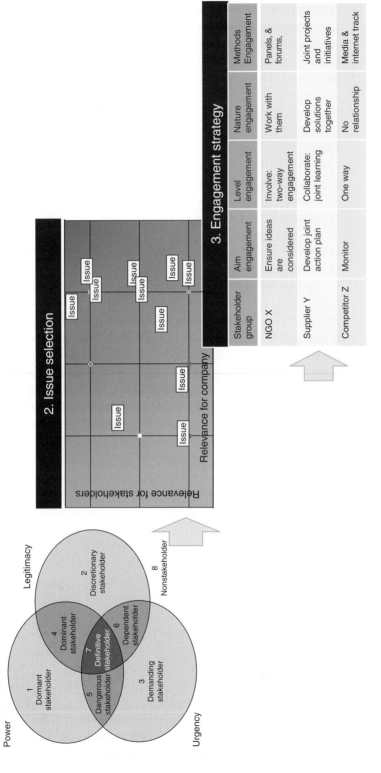

**Figure 7.6 Which stakeholders are important?**

### 7.4.4  *Interaction with stakeholders*

Many studies have shown that the credibility of companies in relation to social issues is not particularly high. Communication is often tackled from the point of view of the company's own interests rather than from the perspective of public interest. Figure 7.7 is based on one of these studies.

NGOs are seen as reliable communicators, representing interests that often coincide with public interests. Combined

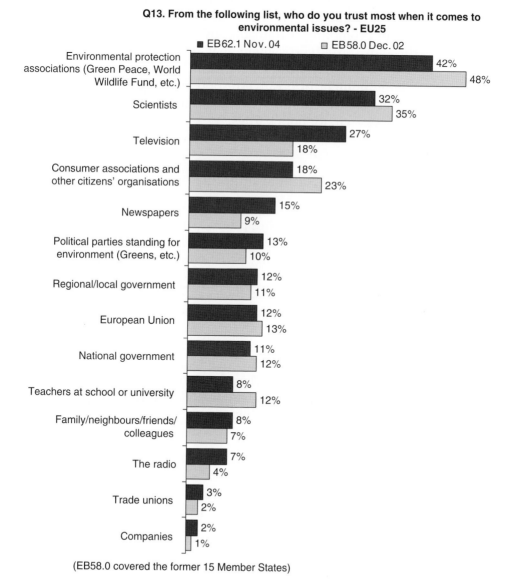

**Figure 7.7   Which sources do European citizens trust for information on the environment?** *Source*: European Commission (2005).

communication between NGOs and companies can therefore strengthen the message of both. The method of communication that is used must be carefully considered, especially by the NGO, since its trustworthiness is at stake.

One example of joint external communication between a dairy brand and an NGO resulted in an advertisement which appeared in a large number of Dutch daily newspapers. In the discussions that preceded their collaboration, the dairy company and the NGO carefully examined how both could convey the message. This joint communication was preceded by many years of cooperation between the two organisations in the interests of promoting sustainable soy. The NGO was able to show through the advertisement that it was working proactively towards its poverty reduction goals. The advertisement prompted a large number of positive responses, with many new donors coming forward for the NGO. So this joint communication was a clear win-win for both parties.

## 7.5   The business case for sustainability

In the long term, making products more sustainable increases shareholder value, since future demand for food to feed 9 billion people by 2050 (FAO forecast[1]) can only be met if it is done in a sustainable way. Sustainably produced food is expected to have more success in the market than alternative products. Moreover, innovation in the sphere of sustainability can help to reduce costs through the use of products and processes which consume less energy and raw materials and generate less waste. This in turn creates value for shareholders, capital providers and society in general. Section 7.5.1 explains the creation of value through increased sustainability. Ultimately this value creation must result in an attractive business case for all parties.

### 7.5.1   *Value creation in making products more sustainable*

Various forms of value creation can be generated by increased sustainability. Certain aspects must be considered in order to properly evaluate the business case for sustainability:

---

[1]   www.fao.org/fileadmin/templates/wsfs/docs/expert_paper/How_to_Feed_the_World_in_2050.pdf.

(1) *Pricing power*. For suppliers, the demonstrable sustainability of their products is a reason to enhance the value of those products. Buyers in turn are prepared to pay more for these products. Organic products are an example, where an increasingly large number of buyers are prepared to pay more for organic than for conventional products.

(2) *Cost savings*. Cost savings can be demonstrated through more efficient consumption of raw materials, energy and water. Costs can be reduced by examining, in cooperation with chain partners, where in the total supply chain efficiency improvements can be made. One example is the drying processes that are very common throughout the dairy sector. The first supplier in the chain supplies a dry bulk commodity (such as milk powder) to the next link in the chain. This producer dissolves the powder in water to formulate it into an end-product. The solution is then dried so that it can be delivered to consumers in its final packaging. By improving cooperation throughout the chain, the first supplier can also organise the entire formulation and remove one drying stage from the process. This substantially reduces the product's carbon footprint, while at the same time cutting energy and water costs. Equal cooperation throughout the chain is necessary to bring about these kinds of measures.

(3) *Motivated employees and increased employee productivity*. Companies that are demonstrably working hard to improve sustainability can attract more highly qualified personnel. Studies have shown that the reputation of companies in relation to sustainability is considered by highly qualified professionals to be a major plus-point. This makes the company an attractive employer. Motivated staff also push down costs, since there is less production line wastage because employees tend to assume more personal responsibility and intervene more readily when there is a production breakdown.

(4) *Improved customer loyalty*. Experience shows that cooperation to achieve greater sustainability equalises relationships throughout the chain. The traditional seller–buyer relationship, in which the only area of negotiation was the price, thus transforms into a partnership relationship, where players work together to make the end-product more sustainable and to jointly use this to add value to the end-product.

This increases customer loyalty. Obviously the parties still negotiate over the price of the product, but there are now many other components in the equation.

(5) *Enhanced ability to enter new markets.* Making products more sustainable opens up new market opportunities, because the company is now offering products in response to the wishes of buyers and consumers who attach great value to the use of sustainable raw materials. The same applies to buyers who do not want to run the risk of using unsustainable raw materials.

(6) *Lower risk premiums.* Banks determine the costs of financing based on the short and long-term risks that are run. Unsustainable processes and products pose a risk for banks, increasing the costs of financing. These costs are passed on to the buyer of the financing in the form of a higher interest rate. Banks are therefore increasingly assessing companies on the basis of their vision and sustainability performances. If a company can demonstrate that it has a specific sustainability policy, the interest rates it is charged tend to be lower than if this were not the case.

(7) *Lower cost of capital through greater access to capital and lower insurance costs.* The costs of capital and insurance depend on the risk being run. As in point 6, banks will include sustainability in the equation when weighing up the risk of providing financing, and also when considering their overall willingness to provide financing. Insurance companies similarly take account of sustainability aspects when calculating insurance premiums, since these affect risk.

In addition to the seven aforementioned aspects, the innovative strength of a company is often also taken into account.

Using these criteria, a demonstration business case can be compiled for making products more sustainable. First it will be necessary to examine which aspects can be directly applied in the business case. Experience shows that these are aspects 1, 2, 4 and 5. The other aspects will lead to cost improvements in the longer term and can be taken into account, but cannot be translated directly into lower costs for the product. The business case is built up by looking at the combined value of all these aspects in the chain. This may involve cooperation with many links in the chain. Managing the chain is therefore vital if the best return is to be obtained. Equality among the partners and confidence in each other's actions is essential. All the partners must jointly establish which sustainability aspects can be used to ensure that the supplier of the end-product adds value. This may involve demonstrating that a specific function has the lowest carbon footprint. All the parties in the chain should then examine how this objective can be achieved, e.g. by reducing the number of drying stages, increasing transport efficiency, cutting sell-by date loss, wastage, etc. These cost savings will be shared by the chain parties who have made the particular contribution. Ultimately this must create a win–win situation for all the links in the chain, generating customer loyalty and ongoing improvements throughout the chain. A large number of demonstration projects have now been launched, which have led to win–win situations for all parties in the chain.

Making more sustainable products is largely down to cooperation throughout the chain. Initiatives taken by key links in the chain will gradually enable all the links to implement sustainability. Cooperation boosts the sustainability of the whole chain, which must then improve throughout.

## 7.6   Policy and strategy adopted by different dairy companies

The appendix to this chapter gives an overview of the sustainability policies, strategies and projects of six dairy companies. This analysis shows that all the major dairy companies have included sustainability in their vision and mission, but the approaches they have adopted are different. Cooperatives clearly opt for a different approach than privately financed companies. Companies with water as well as dairy in their product portfolio (such as Nestlé and Danone) will specifically adopt a sustainable approach to water consumption.

## 7.7   Looking to the future

Clearly, the sustainable production of dairy products is a genuine trend rather than simply hype. Companies that fail to come on board will miss the boat. Society expects companies to use raw materials and inputs carefully and responsibly. Those that fail to meet these expectations will ultimately not be able to rely on their products being accepted. For many companies, there are parallels here with the developments surrounding quality assurance that took place a few years ago. The frontrunners were able to exploit this in their marketing and sales, after which the rest of

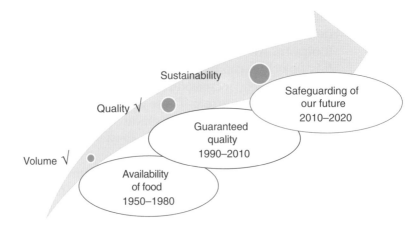

**Figure 7.8 From volume to sustainable food production.**

the market fell into line. The issue of quality altered relationships within the chain. It is clear that these relationships are changing again under pressure from the demand to make products more sustainable (Figure 7.8). Links in the chain will begin to work together to meet this demand. The marketing departments will be first off the mark, in order to position these sustainable products. This will be followed by efforts to define the criteria which these products must meet. Companies must now do much more to render account to society of the sustainability of their products.

Making products more sustainable is an activity that is not just reserved for the major market players. There are also opportunities for the smaller players in the chain. As the key links in the chain begin to move towards a more sustainable approach, this will create opportunities for all the links. Due to the composition of their products, the major players are in many cases dependent on other (often much smaller) links in the chain. By responding in a cost-efficient way to the needs of the larger parties, these smaller companies can exploit the opportunities that have arisen. Due to the speed with which they operate, they can make their own products more sustainable much faster, and can use the marketing strengths of the larger companies.

## 7.8 Conclusion

In recent years, the growing emphasis on sustainability has changed the role of dairy companies within society. Before the turn of the present century, companies were largely responsible

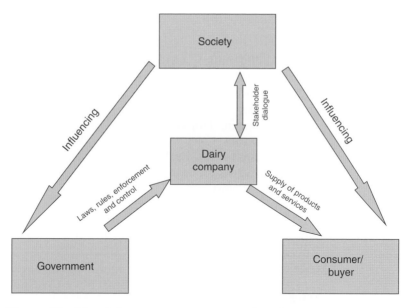

**Figure 7.9    The dairy company as part of society.**

for their own processes. Now, however, that responsibility has increasingly come to be focused on products. This is forcing companies to adopt an entirely new approach. Responsibility in the chain now resides with producers, who have to conclude agreements with all the links in the chain about how to make their products more sustainable. It is not only governments that call companies to account concerning the sustainability of their products; NGOs have also discovered that actively challenging companies about the sustainability of their products works very effectively. Companies with a large brand portfolio are especially susceptible to this approach. The brand represents the high value which has been built up through marketing and communication. Damage to the brand reputation can have major financial repercussions.

Figure 7.9 is a simplified summary of the relationship implications of the various aspects examined in this chapter.

## References

AccountAbility (2011) AA1000 Stakeholder engagement standard. www.accountability.org/images/content/3/6/362/AA1000SES%202010%20print.pdf.

Berns, Maurice, Townend, Andrew, Khayat, Zayna, Balagopal, Balu, Reeves, Martin, Hopkins, Michael S., Kruschwitz, Nina (2009) *Sustainability and Competitive Advantage*, MIT Sloan, Fall 2009.

European Commission (2005) Summary: The attitudes of European citizens towards environment. *Eurobarometer*, no. 213. April 2005. ING Bank. Report.

Lacy, Peter, Cooper, Tim, Hayward, Rob, Neuberger, Lisa (2010) A new era of sustainability. CEO reflections on progress to date, challenges ahead and the impact of the journey towards a sustainable economy. *Accenture*, June 2010.

McKinsey & Company (2010) How companies manage sustainability.

*MIT Sloan* (2009) Survey: The Business of Sustainability Now. *MIT Sloan*, Fall 2009.

Porter, Michael E., Kramer, Mark R. (2006) The link between competitive advantage and corporate social responsibility. *Harvard Business Review*, December 2006.

# Appendix: Overview of the sustainability policy and strategy of various companies (around 2010)

## Unilever

We will develop new ways of doing business with the aim of doubling the size of our company while reducing our environmental impact.

| CSR mission | CSR strategy | Focus areas & activities |
|---|---|---|
| • Creating a better future | • Inspiring billions of people to take small, everyday actions that add up to a big difference | • Halve the environmental foot print of our products<br><br>• Help more than 1 billion people take action to improve their health and well-being<br><br>• Source 100% of our agricultural raw materials sustainably |

## Nestlé S.A.

| CSR mission & objectives | CSR strategy | Focus areas & activities |
|---|---|---|
| • Creating shared value<br><br>• To help advance solutions to the serious problems facing us in water, food, and nutrition security | • Examining the multiple points where we touch society and making very long term investments that both benefit the public and our shareholders<br><br>• To create value for the people in the countries where we are present | • **Environment** - Methane recovery (methanisation). Manufacture, in countries from which we source commodities<br>• **Packaging** - Developing recyclable packaging from renewable resources<br>• **Employee** - Behaviour-based safety (BBS). GLOBE business information system<br>• **Farmer** - Improve the farmers' living standards, environmental practices and water usage<br>• **Product quality** - Quality testing - 60/40+ the largest nutrition-focused programme in F&B<br>• **Child Nutrition** - 'Start Healthy, Stay Healthy'. Nestlé Nutrition Council<br>• **Tackling obesity** - 'Together Let's Prevent Childhood Obesity'<br>• **Society** - Popularly Positioned products (PPPs). Direct store delivery initiative |

**Groupe Danone**

| CSR mission & objectives | CSR strategy | Focus areas & activities |
|---|---|---|
| • Dual economic and social project | • Bringing health through tasty, nutritious and affordable food and beverage products to as many people as possible | • **Environment** - 'Danone Fund for Nature' - investing in wetlands preservation. The Danone, IUCN, Ramsar partnership<br>• **Society** - 'Danone Supporting Life' social programmes. Creating endowment fund for the development of the Danone ecosystem<br>• **Transport** - The Marco Polo Project: from 'all truck' to 'all rail'<br>• **Packaging** - Recycled PET<br>• **Employee** - Danone campus. Learning by Danone health university. Open Sourcing. 30% remuneration linked to social and environmental performance<br>• **Governance** - Diversity. DNA of Groupe Danone<br>• **Developing countries** - Danone for all<br>• **Green Investments** - Invest in socially oriented companies<br>• **Farmers** - Linus Project. DQSE |

**Arla Foods amba**

| CSR mission & objectives | CSR strategy | Focus areas & activities |
|---|---|---|
| • To offer modern consumers milk-based food products that create inspiration, confidence and well-being<br>• To develop our business on a foundation of long term perspectives with respect for, and in harmony with, our surroundings | • From cow to consumer describes our value chain and the process we use to produce and sell our products | • **Agriculture** - The Arlagården quality assurance programme. Profitable whey protein<br>• **Strategy 2013** - Company with the most natural products<br>• **Climate** - To achieve at least a 25% reduction in emissions from transport, production and packaging by 2020<br>• **Product development** - closer to nature. Doubled budget for product development<br>• **Transport** - Investing in new, modern vehicles. Educating drivers. Using renewable fuel<br>• **Employee** - Future 15. The barometer employee survey<br>• **Consumers** - Arla customer care centre. Cook books<br>• **Society** - Launched the Children for Life aid project |

**Dean Foods Company**

Largest processor and distributor of dairy products in the United States.

| CSR mission & objectives | CSR strategy | Focus areas & activities |
|---|---|---|
| • Doing business responsibly and realising that sound business strategies, social responsibility and environmental stewardship go hand-in hand in a successful, sustainable business | • Finding ways to continuously reduce our environmental footprint and improve our business practices | • **Environment** - Reduce dependence on fossil fuels. Invest in methane mitigation strategies<br>• **Packaging** - Switching from Polyethylene Terephthalate (PET) to High Density Polyethylene (HDPE)<br>• **Transport** - New cold plate refrigeration technology. Smart Fleet initiative<br>• **Consumers** - The organic center.<br>• **Farmers** - Horizon Organic Producer Education (HOPE) programme. Use anaerobic digesters<br>• **Society** - Sponsor children's medical centers<br>• **Employees** - Online Healthy Connections programme. Confidential employee assistance programme |

**FrieslandCampina**

**We deliver healthy food, every day**

| CSR mission | CSR strategy | Focus areas & activities |
|---|---|---|
| • To feed the world in a sustainable way | • To achieve climate neutral growth<br>• To seek for a balance between performance needs as a business and commitment to society and the environment. A balance that's sustainable in the long-term, ensuring the continued health of the environment, people and our business | • Health & Nutrition<br>• Care for the environment<br>• Sustainable agriculture<br>• Motivated people<br>• Community involvement |

# 8

# A case study of marketing sustainability

Grietsje Hoekstra,[1] Corine Kroft[2] and Klaas Jan van Calker[3]

[1] CONO Kaasmakers, Westbeemster, The Netherlands
[2] A-Ware Food Group, Lopik, The Netherlands
[3] Sustainability4U, Randwijk, The Netherlands

**Abstract:** A case study is presented on how sustainability can be used effectively as a marketing tool and to add new value to dairy products. Examples of advertising, specialist products and cooperation in the chain are given.

**Keywords:** Beemster cheese, Ben & Jerry's, Caring Dairy, corporate social responsibility, Cow Compass, life cycle analysis, marketing, supply chain, sustainability, sustainable dairy

## 8.1  Introduction

The market for sustainably produced products is growing. An increasing number of consumers, retailers and caterers are interested in high-quality, sustainably produced products. How does a dairy company raise consumer awareness among the public and other potential buyers so that they know that the product is

*Sustainable Dairy Production*, First Edition. Edited by Peter de Jong.
© 2013 John Wiley & Sons, Ltd. Published 2013 by John Wiley & Sons, Ltd.

### Box 8.1   Beemster cheese brand credentials

Beemster cheese is a brand that belongs to the Cono dairy cooperative. The group has nearly 500 members, most of which are based in the province of Noord-Holland. The region's rare blue sea-clay soil makes Beemster cheese unique.

Traditional methods are used to make Beemster cheese. For example, the master cheese makers stir the curds in open draining tubs by hand, and the cheeses are aged naturally.

In 2002, Cono was the first in Europe to offer financial incentives to dairy farmers who allow their cows to roam pastures. The cooperative partnered with Ben & Jerry's ice-cream makers in the Caring Dairy sustainability programme in 2008.

**Beemster cheese: the brand and the beliefs**
- Good cheese can only be made if you invest sufficient time in and pay adequate attention to the preparation process.

- Good cheese can make people happy.

- If you have a really good product, people are prepared to pay more for it.

*Target group*: Consumers who enjoy straightforward, tasty food and who are prepared to pay a little more for it.

**Key message: Beemster makes the difference!**
Because of:

- Caring Dairy

- Eye of the cheese master

- Natural maturation

**Beemster and sustainability**
As a traditional cheese maker, we like to take our responsibilities seriously. Caring Dairy allows us to realise these 'sustainable' ambitions. This is vital because, in addition to taste, quality and its pasture-grazed animals, sustainability is also an important feature of Beemster. Caring Dairy also helps narrow the gap between the dairy farmer and the consumer.

not only delicious and useful but also sustainable? This chapter provides a case study on how sustainability can be transformed into business from a dairy company's perspective.

The dairy company Cono Kaasmakers produces Beemster, a well-known cheese brand. Beemster is an exceptionally tasty cheese made from sustainably produced milk. Beemster is increasing the sustainability of the cow-to-cheese process by virtue of the Caring Dairy sustainability programme, which makes it possible to produce Beemster cheese from sustainable milk. Basically, Beemster is a delicious and sustainable cheese!

This chapter uses the 'Beemster and Caring Dairy' model (see Box 8.1) to show how a dairy company can develop and implement a successful sustainability programme on the farm, in the workplace and in the retail sector. The chapter begins with a definition of sustainable dairy as applied to Beemster cheese. Next, the issues surrounding sustainable dairy are presented, followed by a description of how Beemster puts sustainability into action by organising over 160 workshops for dairy farmers each year, along with other concrete projects. The chapter concludes by explaining how Beemster draws stakeholder attention to its sustainability-related efforts.

## 8.2   What is sustainability?

'Sustainability' is an umbrella term that has quite a few definitions. The variations arise from differences in perception (farmer versus city dweller) and place (varying from country to country), among other things. Most importantly, however, we can conclude that the definitions change over time. The meaning of sustainability is constantly evolving due in part to social and scientific developments (van Calker, 2005).

The United Nations introduced a definition which is widely used today:

> Sustainable development is development that meets the needs of the present without compromising the ability of future generations to meet their own needs. It contains the concept of 'needs', in particular the essential needs of the world's poor, to which overriding priority should be given; and the idea of limitations imposed by the state of technology and social organisation on the environment's ability to meet present and future needs. (Bruntland Commission, 1987)

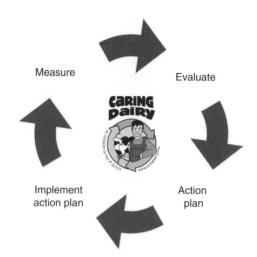

**Figure 8.1   The Caring Dairy improvement cycle.** Courtesy of Ben & Jerry's.

As a sensible, straightforward Noord-Holland cheese maker, Beemster has taken a few words to define sustainability as the balance between the Beemster triple H, namely, 'happy cows, happy farmers and happy planet'.

- *Happy cows.* Our cows provide the milk for the cheese, so animal welfare is an important topic. Healthy cows feel better and live longer.

- *Happy farmers and employees.* Dairy farmers must be able to make a good living from their farming operations. Our employees likewise deserve a good, sustainable workplace.

- *Happy planet.* Attention is paid to the land where our dairy farmers live and work and where our cheese dairy is located. We also look at the impact that our activities have on the climate.

Beemster puts the triple H into practice through the Caring Dairy improvement cycle (Figure 8.1).

Sustainability is dynamic; a process of continuous improvement. The world is changing, and so is the field of sustainability. New information is constantly becoming available. This information is used to advance the Caring Dairy programme and make it even more sustainable.

Beemster uses practical measuring instruments (such as the Cow Compass, section 8.5.2, for example) to gauge progress.

---

**Box 8.2    Ongoing advances in sustainability in the
Konijn family's day-to-day operations**

The Konijn family has been an active member of Caring
Dairy from the beginning. They put the cycle of ongoing
advances in sustainability into practice every day. Here is
an example:

1. Measure: The Konijn family determines how much
   energy they are using.

2. Evaluate: During the 'energy workshop', the family compares their energy consumption to that of other families.
   They discover potential ways to lower consumption.

3. Action plan: Install a heat recovery system as a way to
   reuse the heat generated by the milk-cooling units.

4. Implement: The heat recovery system is installed. The
   heat generated by the milk-cooling units is reused to
   heat the water in the boiler.

5. Measure: Energy consumption has been reduced.

---

The next stage involves reviewing the results and determining
which steps should be taken in order to work more sustainably.
After a certain period, the steps that have been implemented are
measured to see whether they have had an effect and whether
there are new opportunities for taking additional steps. Box 8.2
gives an example of this approach.

### 8.2.1    From cow to cheese

Caring Dairy is not limited to making dairy farms more sustainable. Beemster takes a broader view. After all, making sustainable cheese requires an effort on the part of every single partner
in the chain. Consequently, transport companies and warehouses
are among the other chain partners involved in the process in
addition to the dairy farmers. This is important for several reasons: first, the chain partners' expertise and skills are needed in
order to be able to realise sustainable operations. Second, the
chain partners can be inspired and motivated to join the sustainability movement. Beemster also encourages its suppliers of

goods and services to adopt sustainable working practices. The price–quality ratio is not the only deciding factor when purchasing goods and services; sustainability is taken into consideration, too. This is how Beemster works with its chain partners to achieve a more sustainable chain.

### 8.2.2 Integral approach

Beemster assigns equal importance to the three Hs. In other words, we do not focus exclusively on environmental or animal welfare issues. Each of the three areas is equally important. Accordingly, we always look for sustainable options that benefit cows, farmers and the planet alike. Of course, we are aware of the challenges inherent in this integrated approach. A good example is access to pasture: on the one hand, it promotes animal welfare ('happy cows'). On the other hand, access to pasture can be associated with higher levels of ammonia emission ('happy planet'). It is up to individual dairy farmers to decide which factors take priority as it allows them to make the most sustainable choice for their particular situation, also taking into account the financial implications.

## 8.3 Motivations for sustainability

CSR can be much more than a cost, a constraint, or a charitable deed – it can be a source of opportunity, innovation, and competitive advantage. (Porter & Kramer, 2006)

Organisations do not necessarily share the same motivations for becoming sustainable. For Beemster, the most important motivating factors were ensuring the continuity of the cooperative and the added value for member dairy farms and the Beemster brand.

### 8.3.1 Future licence to operate

Companies are subject to stakeholder demands. Stakeholders include the government, which uses legislation to enforce certain things; social organisations, which draw attention to issues; consumers, who make product demands; shareholders, who want a percentage of the profits; and members, for whom the continuity of the cooperative is a priority. A company must respond to stakeholders' demands and preferences in order to survive on both a short-term and long-term basis. Stakeholders prefer to see companies opt for sustainable operations. For companies, the challenge is to find a method to respond accordingly in a way that ensures long-term success.

According to Willem Lageweg (director of MVO Nederland[1]) and Ruud Galle (director of Nationale Coöperatieve Raad[2]):

> The long-term mindset shared by cooperative enterprises puts 'green sustainability' at the fore. Cooperative enterprises are investing in long-term strategies with the help of their members. For cooperatives, the P in profit has a special – sustainable – meaning. Profit is not the goal; profit is a means. It's what makes investment in the future possible.

Continuity is an important value for Beemster. Dairy farmers established a cooperative over 100 years ago to protect the future of their farms. Working together enabled them to turn perishable milk into cheese that has a much longer shelf life. They also had better access to the market as a cooperative. The Beemster approach to investing in the continuity of the cooperative remains unchanged in the 21st century. Beemster's choice to invest in sustainability makes perfect sense. Responding to stakeholder preferences regarding sustainability ensures that member dairy farmers will receive a fair price for their milk in the future, too.

### 8.3.2  Market opportunities

> Consumers assign more brand equity when efforts to increase sustainability translate into advantages for them. (Niels Willems of Business Openers)

Responding to stakeholder preferences not only matters for the continuity of a company but also presents market opportunities. Consumers are becoming more and more interested in the origin

---

[1]  MVO Nederland (CSR Netherlands) is the national knowledge centre and the national network organisation for corporate social responsibility.
[2]  National Cooperative Council.

of food, animal welfare, the environment and corporate social responsibility (van Calker et al., 2005).

Sustainability is important to 52% of Dutch consumers (Ernst & Young, 2010); 59% of the Netherlands population buys sustainable products. Furthermore, consumers are placing increasing importance on supermarkets carrying organic and sustainable products (Deloitte, 2010). In addition, sustainability can represent added value; it can be a unique selling point (USP) for gaining new clients and reinforcing loyalty among existing customers.

It is possible to can tap into new markets with sustainable products, too. In the food service industry, interest in high-quality, sustainable products is on the rise, which is creating new market opportunities. Beemster cheese was added to Sanday's[3] 'green range', a line of delicious sandwiches with sustainably sourced toppings.

### 8.3.3  Corporate image

Sustainability can help create a positive image. It offers businesses a way to get stakeholders to take notice by communicating the results achieved. Also, some businesses regard corporate social responsibility as a kind of insurance policy; that is, a way to prevent negative publicity and actions on the part of social organisations.

## 8.4  Choose your battles: sustainability strategy issues

Sustainability is an umbrella term. There are many definitions and possibly hundreds of issues involved in the corporate social responsibility question. Social organisations call attention to issues (see the 'Green Saint Nicholas' example in Box 8.3), the government puts topics on the agenda, consumers place demands on products and so forth.

It is up to each business to decide which issues to tackle in its sustainability strategy. This section explains how the various sustainability indicators are formulated in the Caring Dairy programme, followed by a brief discussion of each.

---

[3]  www.sanday.nl.

> **Box 8.3    Non-governmental organisations put issues on the agenda**
>
> Social organisations stimulate companies to adopt sustainable methods. In 2006, Oxfam Novib, the Dutch affiliate of the Oxfam International Confederation, launched the 'Green Saint Nicholas' campaign. In late 2009, the Green Saint called attention to the situation of cacao farmers. The campaign prompted consumers to demand sustainably sourced chocolate letters. In 2009, a mere 15% of all chocolate letters sold in the Netherlands were sustainably sourced; by the end of 2010 the figure had risen to over 95%! Moreover, by the end of 2012 all chocolate letters sold in Dutch supermarkets will be sustainably sourced as a result of the campaign.[4]

### 8.4.1   Determining indicators

Sustainable dairy farming is often compared to organic dairy farming, but this is not entirely accurate. In organic dairy farming, protecting the environment is the guiding principle when making choices. In the Caring Dairy programme, the focus is on achieving the proper balance between protecting the environment (happy planet), animal welfare (happy cows), social progress and economically viable dairy farms (happy farmers). Moreover, the Caring Dairy programme pursues goals (such as reducing the surplus of nitrogen and phosphate), whereas the organic dairy industry follows guidelines (for example, the use of artificial fertiliser is prohibited).

Caring Dairy was launched in Europe in 2002 by ice-cream makers Ben & Jerry's as a pilot in cooperation with eleven dairy farmers, WWF and Wageningen University with the aim to formulate the initial Caring Dairy sustainability indicators. Caring Dairy focuses on a harmonious balance between economic growth, social progress and animal welfare. The programme was structured around the following basic principles:

- high-quality milk with as few additives as possible;
- contribute to soil fertility, air and water quality, landscape and biodiversity;

---

[4]   See www.oxfamnovib.nl.

- make maximum use of renewable sources and minimum use of non-renewable sources;

- sustainable dairy farming benefits the (local) community;

- optimum animal welfare.

Next, an analysis was performed to determine the extent to which the existing initiatives for (sustainable) dairy farming met the above criteria.

### 8.4.2   The basis: 11 indicators

Caring Dairy in Europe currently uses 11 sustainability indicators, all of which are involved in the sustainable production of milk and cheese (Table 8.1).

### 8.4.3   The details: environmental analysis

The indicators in section 8.4.2 are abstract concepts that require further details in order to make them applicable to the cow-to-cheese chain. Therefore, Beemster elaborated on these concepts in cooperation with various stakeholders:

- Dairy farmers: these are the experts with first-hand knowledge of dairy farming-related issues.

- Social developments: what consumers, civil society organisations and government feel is important with respect to sustainable dairy.

- Researchers and experts: informed view and deeper insight into potential opportunities to work more sustainably.

The result is the Beemster triple H: happy cows, happy farmers, happy planet. It is important to set priorities and decide which indicators or issues to tackle as a company. 'CSR policy must relate to your company or product. Therefore, choose themes that you either already have or can have an influence on' (Brüggenwirth & de Ronde, 2009). At Beemster, the focus is on the most important indicators for the Dutch dairy farming industry: animal welfare, fair income, scenic value, energy and climate. The issues are assigned priority in consultation with

**Table 8.1    Caring Dairy's eleven sustainability indicators.**

| | | |
|---|---|---|
|  | *1* | *Animal welfare*<br>Cows that feel good live longer and get sick less often. This benefits both cow and farmer. Taking a critical look at aspects such as food and housing can optimise animal welfare. Encouraging free ranging is likewise important. |
|  | *2* | *Energy and climate*<br>Energy consumption on farms falls into one of two categories: direct (such as electricity, gas and diesel) and indirect (for example, to produce concentrates and artificial fertiliser). Taking a critical look at energy consumption provides insight into possible ways to save. These savings help reduce $CO_2$ emissions, and often lower costs, too. |
|  | *3* | *Biodiversity*<br>Farmers live and work in the countryside. Farms and cows are an essential part of the Dutch landscape. Efforts on the part of farmers helps preserve this landscape and provide flora the opportunity to flourish and avifauna a place to nest. |
|  | *4* | *Social and human capital*<br>The majority of Dutch dairy farms are family businesses without employees. This means that farmers work 7 days a week; after all, cows require daily milking and feeding. Taking a critical look at the work schedule can optimise the balance between work and leisure. Safe working conditions for farmers are part of this, too. |
|  | *5* | *Farm financials*<br>The first matter of importance is that farmers are able to earn a decent living. To this end, it is best if farmers have a clearly formulated strategy and insight into the financial situation of their business. They can use this information to determine whether there are ways to increase the profitability of their operations. |
|  | *6* | *Soil fertility and health*<br>Good, fertile soil is essential for growing grass for cows to graze daily. The more fertile the soil, the more grass will grow. The more grass a farmer has, the less concentrates the cows require. Plus, the farmer does not have to use as much fertiliser. |

*(continued)*

**Table 8.1**    (*continued*)

| | | |
|---|---|---|
|  | 7 | *Nutrients*<br>The use of nutrients (such as nitrogen and phosphate) on a dairy farm must be as high as possible. This is achieved by working with the nutrient cycle. The cycle involves animal, manure, soil and roughage. The challenge is to close this cycle on a farm as much as possible by using inputs (such as concentrates and artificial fertiliser) more efficiently. Closing the cycle has a positive effect on the environment. |
|  | 8 | *Water*<br>Farmers use water for many purposes, including cleaning the milking system and providing the cows with drinking water. Smart use (and reuse) of water can help save consumption, which saves money, too. |
|  | 9 | *Local economy*<br>Bring consumers and farmers closer. Let consumers see and experience what is involved in operating a sustainable dairy farm, for example, by offering guided tours, information on a website, presentations etc. |
|  | 10 | *Pest management*<br>The use of pesticides includes the risk of traces ending up in the water or air. Taking a critical look at the use of these agents can lead to reductions. Possibilities for replacing pesticides with more natural alternatives are investigated, too. |
|  | 11 | *Soil loss*<br>Soil loss refers to the erosion of topsoil as a result of wind or runoff. In the Netherlands, where the soil is largely covered with grass, the problem of erosion is confined to peaty areas. In these regions, methods for using improved drainage to reduce soil erosion are being explored. |

Diagrams courtesy of Ben & Jerry's.

dairy farmers, Caring Dairy advisory board member organisations, and experts and scientists. Of course, attention is paid to the other factors, too; after all, Beemster has an integrated vision of sustainability. Happy cows, happy farmers and happy planet are all equally important. Furthermore the programme is periodically evaluated and harmonised internationally.

Triple H measuring instruments (Table 8.2) have been developed as a way to monitor the steps taken by dairy farmers (see section 8.5).

Based on an environmental analysis, Beemster identified a number of issues related to the sustainable production of milk and cheese. Figure 8.2 shows the issues that Beemster has chosen to explore in further depth, taking account of input from external and internal channels (Box 8.4; Figure 8.3). Several examples can be found in section 8.5.

## 8.5   Getting to work

Environmental analysis provides a framework for the issues that Beemster adopts as part of the Caring Dairy sustainability programme. In doing so, Beemster, together with chain partners, aims to take concrete steps. Various projects will be used to carry out all of the steps. Collectively, the projects are helping Beemster move forward in achieving a more sustainable cow-to-cheese chain.

### 8.5.1   *Happy farmers*

#### Farmers' workshops

Each year Beemster organises over 160 workshops for farmers. Beemster believes that providing farmers with knowledge and skills enables them to increase the sustainability of their operations. The workshops cover a wide range of themes, from animal welfare (for example, udder health or young livestock) to soil management, strategy and agricultural nature conservancy. The workshops are led by experts including researchers from Wageningen University and specially trained veterinarians. Participation is limited to a maximum of 10 dairy farmers and all of the workshops are held on site at dairy farms. This way, the theory covered during the workshop can be directly applied in practice. Participant feedback has been positive. After each workshop, the dairy farmers draw up an action plan outlining the issues that they will investigate further with respect to their specific operations. The workshops actively lead to more

**Table 8.2  How Beemster measures the triple H.**

| | Happy cows | Happy planet | Happy farmers |
|---|---|---|---|
| Indicators | – Animal Welfare<br><br>*Including*<br>– Animal health<br>– Housing facilities<br>– Feed & Water<br>– Young-stock rearing | – Soil<br>– Soil fertility<br>– Water<br>– Nutrients<br>– Biodiversity<br>– Energy/Climate<br>– Pest management | – Local Economy<br>– Farm financials<br>– Human capital |
| Measuring instruments | Cow Compass | Cradle to Cradle (C2C) Compass | Farmer Compass |

Diagrams courtesy of Ben & Jerry's.

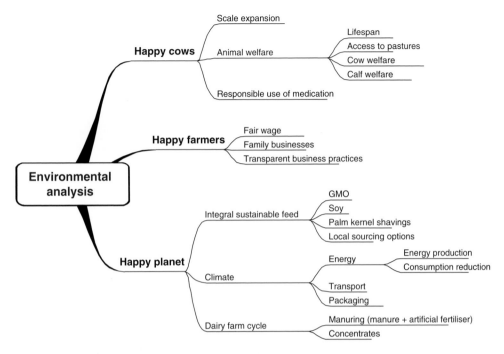

**Figure 8.2   Environmental analysis.**

> **Box 8.4   Sustainability organisation**
>
> Beemster aims to make the cow-to-cheese chain more sustainable. Beemster appointed an advisory board in cooperation with partner Ben & Jerry's (see Figure 8.3). Board members include representatives from various civil society organisations and experts in the field of sustainable dairy. The board plays an evaluating and consulting role to determine whether the Caring Dairy programme is correctly addressing the appropriate sustainability issues. A feedback group of discerning dairy farmers provide input about Caring Dairy in dairy farming operations. Employees from various departments in the sustainability arena develop and implement projects designed to make the cheese-making and transportation processes more sustainable. Another group of employees is responsible for communicating the Caring Dairy programme to stakeholders.

sustainable steps. Examples include adapting the drinking water supply for cattle, and adjusting rations for temporarily dry cows. A questionnaire is distributed annually to monitor progress made on the various issues.

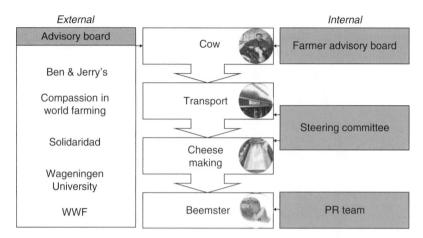

**Figure 8.3 Beemster's consultative body.** Courtesy of Cono Kaasmakers.

### Price premiums for enthusiastic participation

Nearly 500 Beemster dairy farmers have been involved in the Caring Dairy programme. Farmers can join on a voluntary basis. Membership, however, comes with obligations. Participating farmers must meet the following criteria:

- Attend three workshops and draft and implement three action plans each year.

- Use the C2C Compass (section 8.5.3) and Cow Compass (section 8.5.2) to quantify sustainability performances. If farmers receive unsatisfactory marks for certain indicators, together with an employee they will look for opportunities for improvement.

Farmers who meet the above criteria will receive a Caring Dairy price premium of €0.50 per 100 kg of milk.

### Active employees

Beemster employees are encouraged to think about sustainable solutions for the workplace and at home. They are invited to offer suggestions for making transportation and cheese making more sustainable. Together with experts, employees look for the most sustainable solutions with respect to the layout of the cheese dairy, both inside and outside.

Employees are stimulated to adopt sustainable practices at home, too. They are eligible for special leave in exchange for their commitment to social or environmental sustainability in the form of 'volunteer days', which they can use to spend time working for a good cause. Employees can also request sponsorship for sustainable activities. Several employees have participated in Roparun, a non-stop relay race from Paris to Rotterdam that raises money for people with cancer. Beemster provided the runners with financial support.

### 8.5.2   Happy cows

Thanks to nearly 30,000 cows, Beemster receives over 300 million kilograms of milk each year. This milk is turned into Beemster and other varieties of cheese. Our farmers take very good care of the cows that provide the outstanding quality milk for our exceptional Beemster cheese.

### Free-range cows

Beemster began encouraging its farmers to let their cows graze in the fields from spring until autumn in 2002. Farmers receive a price premium of €0.50 per 100 kg of milk when they give their cows access to pasture. This partly explains why approximately 95% of the Beemster dairy farms do this. The national average is less than 80% (Central Bureau for Statistics, 2010[5]). We believe that access to pasture positively affects the welfare of dairy cows. Furthermore our experience shows that milk from cattle that eat grass has a different composition. Every year, Beemster makes Graskaas or 'grass cheese' from the first batch of milk produced by the grazing cows. Lastly, Beemster believes that cows belong in the human-made Dutch landscape. This position is championed by various civil society organisations that subscribe to the importance of grazing. According to Dierenbescherming, the Dutch animal welfare organisation, cows are not meant to remain in a shed. Being outdoors in a pasture allows them to express their natural behaviour to the fullest (Dierenbescherming, 2011). The Dutch animal rights organisation, Wakker Dier, campaigns to encourage grazing for cows.

---

[5]   http://statline.cbs.nl/StatWeb/publication/?DM=SLNL&PA=7073 6ned&D1=0-5&D2=a&D3=4-6,l&VW=T

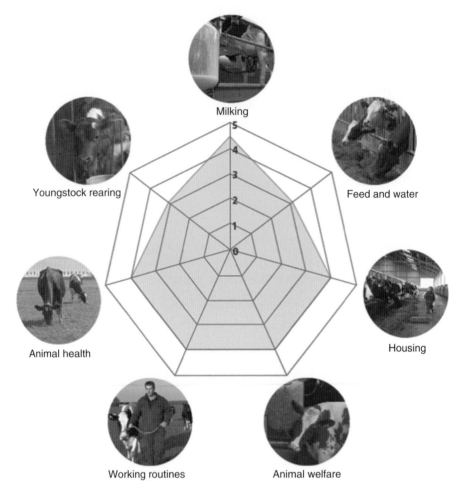

**Figure 8.4    Cow Compass.** Courtesy of Cono Kaasmakers.

## Cow Compass

How can we tell whether a cow is happy? Beemster has worked with dairy farmers and veterinarians to develop the Cow Compass (Figure 8.4), a tracking system for dairy farmers to gauge the health and happiness of their cows. In addition, the compass provides insight into what dairy farmers can do to improve the welfare of their cows. It is a practical tool appreciated by many dairy farmers. Based on 40 performance indicators, farmers gain insight into seven critical success factors that affect animal health and welfare: milking, food and water, housing, animal welfare, youngstock rearing, work routine and animal health. The Cow Compass offers farmers direction for additionally increasing animal welfare. Plus, farmers can use the Cow Compass to show the outside world how they look after the health and welfare of their cows in their dairy operations.

### Feed

Beemster aims to increase the sustainability of the cow-to-cheese chain, and cattle feed plays an important role. A cow's diet largely consists of grass. The menu is often supplemented with concentrates or dried food, which frequently includes soy or palm kernel shavings. There are downsides to the production of raw materials. With the support of Solidaridad and Natuur & Milieu, Beemster is investing in sustainable cultivation methods for these particular raw materials. Support is given to local growers, for example by organising workshops, to help them adopt more sustainable practices. Key topics include preventing deforestation, the use of pesticides and working conditions.

### Good Dairy Award

In 2011, Beemster and its dairy farmers' commitment to animal welfare earned the Good Dairy Award. The international animal welfare organisation, Compassion in World Farming, presents the award to businesses in recognition of efforts to promote grazing, optimum welfare of calves and cows, and other related aspects.

## 8.5.3  Happy planet

Three-quarters of the land in the Netherlands is occupied and used by farmers. The challenge facing dairy farmers is to make the most of what the Earth has to offer. This allows them to produce the same amount of milk with fewer inputs such as concentrates and artificial fertiliser.

### C2C Compass

Dairy farmers can use the C2C Compass (Figure 8.5) to measure the happiness of the planet. Six critical success factors help farmers see how they fare in terms of minimising the

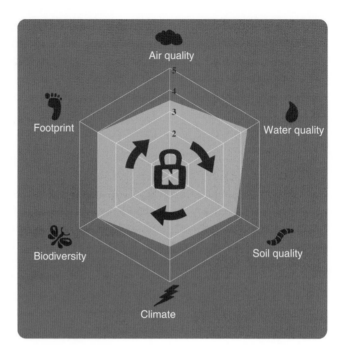

**Figure 8.5   C2C Compass.** Courtesy of Cono Kaasmakers.

environmental impact of their farms. The more contained the cycle, the more effective their use of the Earth. One indicator is that they do not need as much artificial fertiliser or concentrates, which is better for the planet and has a positive impact on the farm's profits, too.

The C2C Compass assesses the following factors: air quality, water quality, soil quality, climate, biodiversity and footprint. Beemster uses two variations of this compass. The C2C Compass[basic] was developed by the Netherlands Centre for Agriculture and Environment. Using a questionnaire, it provides dairy farmers with a general overview. Dairy farmers who wish to gain a more detailed view of the cycle can use the C2C Compass[plus], which comprises a questionnaire and indicators. The Plus version was developed by Duurzaam Boer Blijven and corresponds to the *Kringloopwijzer* developed by researchers at Wageningen University.

**Life cycle analysis**

Dairy farms use the C2C Compass. In order to gain insight into the impact of the entire cow-to-cheese chain, a research bureau put together a life cycle analysis to shed light on the environmental impact of cheese production. The analysis looks at variables such as phosphate emissions, land use and energy. For now, water is left out of the analysis due to the lack of internationally recognised calculation methods.

### Sustainable transport

In 2010, the efforts on the part of Beemster and its carriers were rewarded with the Lean & Green quality mark, which is awarded to companies that successfully reduce their $CO_2$ emissions by at least 20% within five years. Beemster managed to do so by implementing various measures, such as installing dual-fuel systems in a number of the milk trucks. The combined diesel and gas engine reduced $CO_2$ emission by 6% and reduced soot emission by 10%. In addition, drivers were instructed on *'het nieuwe rijden'*, a new approach to driving that promotes maximum efficiency, and routes were optimised.

### Our ambitions

Beemster is not a passive organisation; after all, sustainability is a continuous process. Our plans for the future include:

- Making a critical assessment of our environment and integrating new issues in our sustainability programme as necessary.

- Integrated sustainable feed: instead of focusing on more sustainable alternatives to individual ingredients such as soy and palm kernel shavings, we are working with stakeholders to assess the feasibility of an integrated sustainable feed that would respond to multiple sustainable issues, including $CO_2$ emissions and raw ingredients such as soy and palm.

- In cooperation with our dairy farmers we are investigating the feasibility of having the farms generate green energy (e.g. using solar panels, bio-fermentation and windmills) that Cono Kaasmakers, Beemster's parent company, in turn can purchase from them.

- Beemster wants to share aspects of the sustainability programme with others. This will provide additional dairy farmers with the opportunity to make their operations more sustainable. The Cow Compass is a good example: a number of other farmers involved in the dairy industry have begun using the compass.

The path that Beemster chose to make the cow-to-cheese chain more sustainable starts with a careful analysis and action plan, followed by concrete steps. In other words, there is more than just talk. Action is taken. To this end, Beemster strives to work closely with chain partners including dairy farmers, carriers and civil society organisations. As a result, Beemster develops many projects to increase the sustainability of the cow-to-cheese chain. Collectively, these projects enable Beemster to make considerable progress in producing cheese that is even more sustainable.

## 8.6  Communicating sustainability

Sustainable business operations play a role in the image and continuity of a dairy organisation. Sustainability also offers opportunities vis-à-vis customers. You can choose a proactive approach to communicating your sustainability-related steps to stakeholders. Your method of communication depends on your chosen strategy: if sustainability is part of brand value, then your emphasis will be on marketing communication. If it is primarily intended to positively influencing your image, then corporate communication is the obvious choice.

Basically, we can take either a proactive or a reactive approach to communicating about sustainability. Reactive is geared to preventing problems with stakeholders. Proactive provides an opportunity to work on a brand or corporate image (Brüggenwirth & de Ronde, 2009).

The four Ps in the marketing mix are helpful when marketing sustainable products.

1. *Product*. Sustainability can be one of your product's brand values. Sustainable practices can change the flavour of a product. For example, every year Beemster produces the limited edition Graskaas using the first batch of milk produced by the cows when they return to the pastures in springtime. This 'grass cheese' is prized for its mild, tangy flavour.

2. *Price*. Investing in sustainability generally costs money. Part of the investment will be recovered within several years (for

example, as a result of lower expenses). Some of it will result in a higher cost price, which may be passed on to the end-user.

3.  *Place*. The place where the product is sold – is it on the bottom shelf or featured with other sustainable products in the supermarket? Putting your product in 'sustainable shops' such as Marqt adds to your image, too. For example, the Beemster roll is part of Sanday's 'green range', a line of sandwiches with sustainably sourced toppings.

4.  *Promotion*. Sustainable products offer the opportunity to distinguish yourself from other products on the market. In addition to flavour and quality, Beemster also stands apart from other cheeses because it is made from sustainably produced milk.

Basically, there is plenty of potential for making the most of sustainability when marketing your products. This section provides useful suggestions for incorporating sustainability in your marketing and communication strategies.

### 8.6.1  *Message*

Communicating a sustainable message is often a challenge. In many cases it concerns technical aspects that are not readily associated with product advantages. Cheese does not automatically taste better because of sustainably produced milk, but it does make consumers feel better. After all, the milk is produced by cows whose welfare is optimally protected. Plus, attention is paid to the environment. Two caveats apply to all sustainable messages: 'Do what you say' and 'What you say must be true.'

An open and honest attitude is crucial. Share your beliefs and explain your priorities (see Box 8.5). Of course concrete figures and end results matter, but the path to increased sustainability is just as interesting. Show what you experienced during your journey to achieve more sustainability, and the obstacles and triumphs along the way. Take your target group along on your trip. Consumers do not want stylised pictures; they see right through them (Brüggenwirth & de Ronde, 2009). A product must have credibility, and be a natural extension of the organisation's true identity. The message that businesses communicate must be authentic. Authenticity is what appeals to consumers (van Eck et al., 2008). Finally, make sustainability relevant to

> ## Box 8.5 Integrated communication
>
> It is not uncommon for communication, corporate, internal and marketing to be kept separate in practice. Van Eck, Willems and Leenhouts (2008) even refer to a 'Great Wall' that exists between marketing and corporate communication. With regard to sustainability, however, the messages shared by the various disciplines have a lot in common. Therefore it is wise to effectively combine these disciplines and present an unambiguous message. Successful communication starts with integrated communication. This applies to every theme, especially sustainability!

your target group. Make sure that your target group understands your message. The sustainable message must prompt recognition among the target group (van Eck et al., 2008). The challenge is always to find an appropriate balance between sustainability and the other brand values (Brüggenwirth & de Ronde, 2009).

### 8.6.2 Means

Many books and articles have been published about the means for executing a marketing strategy. This section explains the means that Beemster has chosen to draw stakeholder attention to sustainability.

### Partnerships

Civil society organisations have a great deal of expertise, which you can use to shape your sustainability programme. In addition, you can support social organisations by spreading your sustainable message. Claiming that you are the best is neither effective nor appealing (van Eck et al., 2008). It increases the credibility of your message when social organisations share your sustainable steps.

For example, in 2010 Natuur & Milieu endorsed the introduction of Beemster Graskaas. In addition, Beemster and Solidaridad work closely on sustainable livestock transport. As a result, the Solidaridad website encourages consumers to buy Beemster cheese because Beemster invests in sustainable transport.

## Quality mark or label

There are many different labels to indicate that a product has been produced sustainably. The advantage of adopting an existing label is that consumers are likely to be familiar with it. But be sure that the logo reflects your activities, and avoid crowding your packaging with too many labels. This way, you will ensure that consumers know which message applies to your product, as opposed to confusing them with an overload of labels.

For Beemster and Ben & Jerry's, there was no existing certification scheme on the market that accurately illustrated what the Caring Dairy initiative was all about. Consequently, the partners formulated their own standards and designed their own logo, which is featured on Beemster and Ben & Jerry's packaging. It is the quality mark for all activities associated with the Caring Dairy programme.

## Reporting

The saying goes that there is power in numbers, which is true for sustainable communication. Third parties, such as customers and social organisations, want to know how sustainability efforts are paying off. Is progress actually being made? A sustainability report is a useful tool. You can choose between producing a separate social responsibility annual report or incorporating it in your regular annual report. List your achievements and ambitions. Transparency is key. In other words, be honest about failures. You gain far more credibility through honest and open communication as opposed to painting an overly optimistic picture of your sustainability-related activities!

## Public relations

Each year, Beemster organises a Beemster Sustainability Day. The event brings sustainability and corporate communication managers from retailers and caterers together to learn about the latest developments in sustainable business practices. Other associates including social organisations, research institutions and government officials are invited, too. Beemster's objective is to host an event with a broad focus that provides guests with information about the steps that Beemster has taken, and gives other stakeholders an opportunity to share their vision as well.

In so doing, Beemster creates an annual platform for sharing knowledge and expertise in the field of sustainable food.

Beemster organises meetings with stakeholders as necessary. For example, dairy farmers are invited to learn about the steps that Beemster is taking.

Awards are regularly given to businesses committed to sustainable practices. In 2010, Beemster was named as winner of the Boerderij Business Top 100 in the sustainability category. As already mentioned, in 2011, Compassion in World Farming chose Beemster as the recipient of the Good Dairy Award. These awards are an independent confirmation of Beemster's sustainability efforts. They also offer good opportunities for publicity, for example, in a joint press release.

### Personal contact

Doing business sustainably means networking, not only to establish contacts to help you continue to develop your sustainability strategy but also to promote your sustainability approach. Investment in your network is crucial. Make appointments and give presentations about sustainable business practices during conferences (and attend conferences, too) to expand your network and increase your knowledge. Personal conversations with customers are valuable, too, and offer insight into what is important to them. You can use this information in your programme to create market potential.

In brief, invest in your network and listen to your stakeholders. Finding out what your stakeholders consider important will allow you to further develop your sustainability strategy and tailor your communication about it more effectively.

### Website

Usually, there is limited space on packaging to provide more than the basic outline of your sustainability programme. For consumers whose interest has been aroused by your packaging, a website is an ideal place to provide background information.

### Social media

More and more people are actively using social media such as Facebook, Hyves and Twitter. You can use these means of mass communication, too. Share what your company is focusing on in terms of sustainable activities; create a fan base of followers. As a company you can use social media to follow key opinion leaders in the field of sustainability and incorporate their feedback as you continue to develop your sustainability strategy.

### Press contacts

Keep the media informed about your sustainability-related activities. Focus on journalists who report specifically on sustainability or farming. Make sure that they have a file. The idea is that they automatically think of your company whenever they sit down to write an article about sustainable business practices.

### Ambassadors

Nearly 500 dairy farmers supply the milk for Beemster cheese. These dairy farmers are the ambassadors of the Caring Dairy programme. They are credible, authentic Caring Dairy representatives; after all, they are the people who put sustainability into practice on their dairy farms. Beemster works with dairy farmers who are interested in talking about Caring Dairy, for example by providing information and cheese samples for open house events and guided tours. At the cooperative's request, dairy farmers also open their doors to visitors such as journalists and government officials who are interested in learning more about the Caring Dairy programme.

## 8.7  Conclusion

Caring Dairy is how Beemster works toward increasing the sustainability of the cow-to-cheese chain, which involves many issues. Milk is the most important ingredient in cheese, which is why Beemster focuses on the issues that are relevant to the Dutch dairy industry. They do so under the motto 'happy cows, happy farmers, happy planet'. The three Hs also occupy centre stage in communication to stakeholders about Beemster's sustainability-related activities. Communication revolves around a clear message (do what you say), openness, honesty and transparency (sharing achievements and points to improve). Beemster shares this message through a wide range of channels, including a website, social organisations and in person. Beemster is increasingly integrating sustainability in its marketing strategy. This makes Beemster very special. It is a very tasty cheese. And a cheese that is produced in a very special and sustainable way. Clearly, Beemster is serious about sustainability.

## Acknowledgement

During the writing of this chapter Corine Kroft and Klaas Jan van Calker (father of the Caring Dairy programme) were with Cono Kaasmakers, Westbeemster, the Netherlands.

# References

Brüggenwirth, B., de Ronde, Anick (2009) MVO en marketing: gaat dat samen? http://nima.nl/inspiration/nima-netwerk/nima-marketing-mvo.

Brundtland Commission (1987) *Our Common Future, Report of the World Commission on Environment and Development.* Published as Annex to United Nations General Assembly document A/42/427, Development and International Co-operation: Environment. 2 August 1987.

Deloitte (2010) *Consumentenonderzoek 2010.* http://www.consultancy.nl/nieuws/deloitte-analyse-van-de-consumentenmarkt-in-2010.

Dierenbescherming (2011) Welzijnsproblemen melkkoeien. August 2011. http://www.dierenbescherming.nl/welzijnsproblemen-melk-koeien.

Ernst & Young (2010) *Duurzaamheid in de Aanbieding: Kansen voor maatschappelijk verantwoord ondernemen voor retailers en hun leveranciers.* http://www.ey.com/NL/nl/Newsroom/PR-activities/Duurzaam heid-in-de-Aanbieding_verslag.

NEN (Netherlands standards organisation) (n.d.) www.nen.nl/web/MVO-ISO-26000.htm.

Porter, M.E., Cramer, M.R. (2006) Strategy and society: the link between competitive advantage and corporate social responsibility. *Harvard Business Review*, December 2006.

van Calker, K.J. (2005) Sustainability of Dutch dairy farming systems: a modelling approach. PhD thesis. Wageningen University. [Summary in Dutch.]

van Calker, K.J., Hooch Antink, R.H.J., Beldman, A.C.G., Mauser, A. (2005) Caring Dairy: a sustainable dairy farming initiative in Europe. 15th Congress IFMA, Brazil, pp. 81–88.

van Eck, M., Willems, N., Leenhouts, E. (2008) *Internal branding in de praktijk: het merk als kompas.* 4th edition. Pearson Education Benelux.

# 9

# Cradle to Cradle for innovations in the dairy industry

Wil A.M. Duivenvoorden

Royal Haskoning, Rotterdam, The Netherlands

**Abstract:** Achieving an innovative and sustainable design fora new dairy plant with a multidisciplinary team requires a systematic approach to the design challenges from the start, thus facilitating optimal communication between all team members and securing an efficient iterative decision-making process as well. By analysing the total value chain of dairy according to the C2C® philosophy, the conceptual design of the plant can be inspired in the most optimal way to meet future demands for sustainability and effectiveness on the part of all stakeholders. With integral design in dairy via the C2C® principles, in fact sustainability becomes a dominant design issue.

**Keywords:** Cradle to Cradle, C2C®, dairy plant, integral multidisciplinary design, sustainability

*Sustainable Dairy Production*, First Edition. Edited by Peter de Jong.
© 2013 John Wiley & Sons, Ltd. Published 2013 by John Wiley & Sons, Ltd.

## 9.1    Introduction

During the 1990s the principles of Cradle to Cradle (C2C®) were introduced and developed by Michael Braungart, William McDonough and colleagues as a paradigm-changing platform for innovations, which leads to a positive footprint instead of a focus on design simply to improve efficiency and quality. Their visionary approach challenges every architect or engineer to redesign their buildings, products and processes in a radical way. For each design these principles can inspire an innovative approach when, according to C2C®, analysis of the so-called biological cycle and the technical cycle (see Figure 9.1) are used in an integral way to guide the total (hence multidisciplinary) design.

C2C® offers the 'proper' spectacles (Braungart & McDonough, 2008) through which to view a design demand beyond the efficiency approach which is normally the goal when aiming to

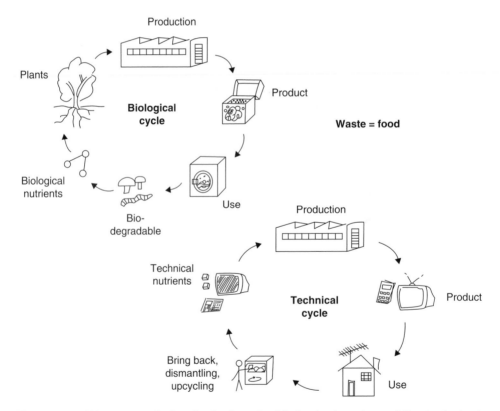

**Figure 9.1    Waste equals food: closing the biological cycle and the technical cycle.** *Source*: Royal Haskoning, Corporate vision of Royal Haskoning inspired by Cradle to Cradle®, 2010.

reduce carbon footprint; C2C® involves effectiveness measures in energy as well as in materials consumption and helps to avoid wastage of both by stimulating upcycling and reuse.

In this way C2C® is much more than just reducing the carbon footprint; instead of simply trying to achieve reduction of this negative footprint, it strives for positive footprints, thus being more effective in the chain of production and consumption of goods.

### 9.1.1  Three main principles

The three principles of C2C® are:

(1)  waste equals food (waste = food);

(2)  use of current solar income;

(3)  celebrate diversity.

Implementation of these principles at the start of the design process, as guidance for interdisciplinary communication and cooperation between all specialists involved, will secure effectiveness as well as sustainability in the conceptual design from the beginning.

As a conceptual framework the principles have already been applied in several designs and processes, from the redesign of cities (e.g. Almere in the Netherlands) to eco-effective carpet manufacture by Desso. Putting the project design's demands and aspirations within a framework of the C2C® principles will lead to innovative, effective, consistent and sustainable solutions to fulfil the demands of modern consumers and society. Basically, according to this approach, the design then aims at 'only doing the right things' instead of 'doing things right', as well as at 'increasing added value' instead of 'reducing negative impacts'.

We now look more closely at each of the three C2C® principles.

#### Waste equals food (waste = food)

In the design of all human production processes, both the technical cycle and the biological cycle could be closed by reusing all 'nutrients' in these cycles for something else. Then there would be no waste any more and the industrial model shifts from 'ownership' to 'usership'.

According to this approach one views all materials as 'nutrients' that should permanently circulate in their own safe and healthy cycles, where after each use they have a new application, and thus are never lost, wasted or devalued.

This implies that under all circumstances the biological cycle and the technical cycle need to stay separate, in order to avoid poisoning each other or obstructing recovery in the other cycle.

### Use of current solar income

The sun is the origin of all energy on Earth and in fact has been supplying to the Earth, day in and day out, for countless ages, more energy than it needed. In a balanced system this energy has always been reflected, absorbed, stored or consumed by the ecosystem of Earth; the future challenge for humankind is to keep this energy system of our Earth in balance while human population continues to increase.

If the growing population of the Earth continues to use fossil fuels (a form of ancient stored and concentrated solar energy) for their energy needs in the same way as modern society does now, this will introduce a large and unacceptable imbalance to the ecosystem of Earth.

Current and future energy demands therefore need to be satisfied in a more balanced way, using the sun as a primary source, i.e. looking for solutions that employ renewable energy sources at all stages of the life cycle of the products. Energy solutions (e.g. electricity, heating and cooling) should always be 'clean', without toxic or other harmful characteristics.

### Celebrate diversity

In the natural world ecosystems are in balance with each other; because of the high level of diversity the systems keep themselves strong and sustainable, though being dependent on each other. For industry this principle means promoting and combining diversity in all functions: biological, social, technical, cultural, and so on. For companies working in industry this means respecting diversity by choosing to use local materials and energy and adapting to local conditions and developments (cultures, societies and landscapes, etc.).

### 9.1.2  *Terminology*

The Cradle to Cradle concept uses some specific terms:

*C2C principles*: Guiding principles of the C2C business concept.

*C2C spectacles*: The lens of the process designer through which he looks at design challenges and problems focusing on sustainable business.

*C2C elements*: New (design) concepts developed by a team of people looking through C2C spectacles.

*C2C roadmap*: The roadmap built on C2C elements to achieve a C2C process design.

*C2C projects*: Projects in which C2C elements are realised.

### 9.1.3  *C2C® and food production design*

For the food industry C2C® can inspire at the start of the design of industrial plants. The first step is to analyse and create an overview of the total value chain, before defining the hierarchical design points in a Program of Requirements (POR). Starting the conceptual design of a plant in this way offers opportunities for eco-effective and sustainable innovation, because up until 2012 most products in the food and beverage industry carry a heavy $CO_2$ footprint and heavy water footprint and are mostly far from closing their biological and technical cycles!

Almost all modern food products are produced in very energy-intensive processing systems, and mostly still use non-biodegradable packaging materials to enable handling, storage, transport and display of the products on their way to the consumer. Furthermore, for a large part of our total western food consumption the route from factory to consumer (and from the farmer to the factory!) is often 'a long way', requiring long-distance transport and/or cold storage to safeguard product quality at all stages. This method of food production ignores both cycles and creates a one-way transport of valuable materials, as norm in which high energy consumption and unacceptable environmental impacts are still accepted.

The C2C® concept, however, requires doing the right things from scratch and puts the use of resources and energy for production into a life cycle perspective, including (re)use and recovery. The major aspiration for food production, according to C2C®, should be the creation of products with only a positive impact on all related (human) activities, so with an energy-neutral or even energy-positive footprint, and with the technical and biological cycles fully closed. In this way it differs from the regular approach of industrial food production, in which efficiency and cost effectiveness take priority, but may be guided by governmental restrictions regarding environmental protection or energy-saving goals.

Of course, in practice, design according to C2C® is not always directly 100% applicable. Equally, the absolute closing of the biological and the technical cycle may not be 100% achievable. However, this should not prevent the start of a C2C® (re)design process when approaching a specific food chain; setting SMART goals as the aspiration for the near future in line with the company's vision and strategy will challenge all stakeholders to come up with innovative opportunities in this chain. So with 'C2C® spectacles' on, one could achieve these SMART goals, by conceiving innovative projects

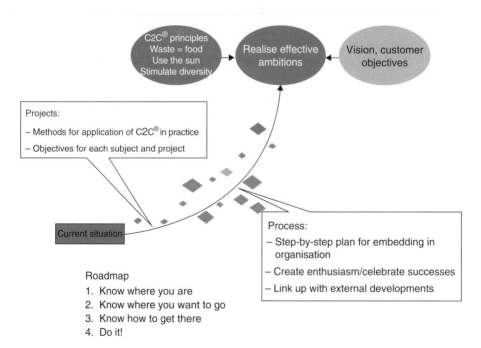

**Figure 9.2   Roadmap to effective aspirations by application of C2C® elements.** *Source*: Royal Haskoning, *Visie Royal Haskoning: duurzame gebouwde omgeving*, September 2010.

and implementing these projects singly as 'C2C® elements in a C2C® roadmap'. The roadmap enables positioning of these C2C® elements as specific projects, en route to achieving the ultimate objectives at the end of the roadmap in a step-by-step manner (see Figure 9.2).

### 9.1.4  A fourth C2C® principle

Understanding the implications of choosing a design process according to C2C® principles reveals that the key success factor in securing results in the total value chain is partnerships. In fact partnership could be called the fourth C2C® principle, as it is a leverage mechanism within chains for successful implementation of effective innovations in design.

This applies to waste, energy, packaging, logistics, water and carbon footprints, human capital, etc., all of which can be analysed according to the three C2C® principles in each value chain.

### 9.1.5  The sustainability matrix

The sustainability matrix in Table 9.1 is an example of how to obtain an overview of all items to be discussed, when seeking the values of sustainability according to C2C® principles. Table 9.1 shows a model for interrogating several aspects (energy, materials, ecosystems, water and human capital) on relevant values such as ecology, economy and fairness, and offers a view on a way to solutions, with partners, for the associated problems. In fact it makes the connection between ambitions and the assessment framework of the various aspects for the project design, based on the values of ecology, economy and fairness. The model can be adapted to every situation and applied at various scales in setting goals for the end of the company's roadmap to C2C® ambitions.

Taking the example of water in Table 9.1, the model shows that from an ecological point of view the production of contaminated water is a serious problem for a food production plant. A solution is possible based on decentralised sanitation with a water treatment process using algae, and business-wise (life cycle costs) this solution also appears to be acceptable, enabling a tick on 'Clean', 'Cleaner discharge than intake' and 'Water footprint', which is a positive result.

Table 9.1  Sustainability matrix for challenging C2C® aspirations.

| | | VALUES | | | | | | | |
| | | Challenges / solutions / ambitions | | | | | Checking | | Fairness |
| | | Ecology | | | | Economy | | | |
| Aspects | Nature | Biodiversity | Health and safety | Climate change | Scarcity | Costs/benefits | PR | Corporate social responsibility | Honesty |
|---|---|---|---|---|---|---|---|---|---|
| Energy | Problems | SO₂ / Acid rain | NOₓ/PM | CO₂/GHG | Fossil fuel | Pay-back time | 'Net positive' | Net energy positive buildings | Super energygrid |
| | Solutions | Solar energy, wind, environment, geothermal energy & biomass, nuclear energy | | | | | | | |
| Materials | Problems | Exposure to waste* | Exposure to air quality | GHG | Natural material | Life cycle analysis | 'Recycled' | Self-cleaning buildings | Leasing of materials |
| | Solutions | Non-PBT, non-CMR, from downcycling to upcycling, materials based upon biomass | | | | | | | |
| Eco systems | Problems | Contaminated water | GMOs (?) | Adaptation, erosion | Phosphate, regeneration, agriculture | Total cost of ownership | 'Green' | Positive contribution to ecosystem | Ecotax |
| | Solutions | Closed biocycles, vegetarian strategies, management of organic soil | | | | | | | |
| Water | Problems | Contaminated water | Contaminated water | Flooding | Fresh water, drought | Life cycle costs | 'Clean' | Cleaner discharge than intake | Water footprint |
| | Solutions | Algae, zeolites, membrane technology, nutrition regeneration, natural water storage, decentralised sanitation | | | | | | | |
| Human capital | Problems | Loss & degradation, contamination | Allergies, infections | Adaptation | Food security | Direct and indirect costs & benefits | 'Health' | Strengthening | Fairtrade |
| | Solutions | Health programmes, (industrial) hygiene programmes, microcredits, food programmes | | | | | | | |

*Toxic, carcinogenic or mutagenic Challenges / solutions / ambitions and checking model, 15 April 2010.
Source: Royal Haskoning, Corporate vision of Royal Haskoning inspired by Cradle to Cradle®, 2008.

## 9.2    A tool for C2C®-driven innovation (PROPER model)

### 9.2.1    Description of the PROPER model

Because of the necessary focus on energy and mass flows in a complex and diverse environment, designing according to C2C® requires a conceptual design approach.

As an aid in organising the conceptual design in such a way that the correct problems are identified (instead of solving problems correctly) the Delft Systems Approach offers the PROPER model (Veeke et al., 2008), in which organisation and transformation of information are achieved in a powerful way via a 'black box' approach.

Because of this black box approach, the PROPER model fits extremely well to the challenges associated with the multidisciplinary design or improvement of a processing plant and its environment, according to the C2C® business concept.

In principle this model (see Figure 9.3) considers the whole plant as well as its constituent processes as black boxes, thus defining the plant and its processes as 'operational systems' in terms of performances and processes.

In this approach the plant and each of its processes are seen as 'PROcess-PERformance' models (hence a 'PROPER' model). Both the control flow (performance) and the product (process) interact with the environment (see the incoming and outgoing arrows in Figure 9.3). From the start of the conceptual design, whatever the discipline, only the 'what' (process) and the 'why'

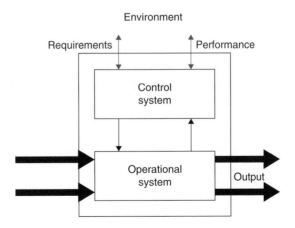

**Figure 9.3    Multidisciplinary design of a system with the PROPER model.** *Source*: Veeke et al. (2008). Reproduced with kind permission from Springer Science + Business Media B.V.

(performance) are defined (and not 'who', 'how' and 'when'), and of course the impacts or exchange with the environment of the process. In this way all the disciplines and specialists involved are given complete freedom with respect to physical interpretation.

According to Veeke, a function description must be determined through abstraction from physical reality in order to construct a conceptual model. 'How' something is done is not important, but 'what' is done and 'why'. This offers two advantages:

(1) It stimulates creativity and allows radical change in the concepts to be realised, either through the use of different technological tools and equipment or by combining functions in a different way. It is a structured route to innovation, as opposed to the other path, 'accidental creation'.

(2) If the design is recorded in terms of functions, the basic assumptions and choices made during the design process remain clear and accessible for future (re)design projects. This construction of 'memory' prevents the 'reinvention of the wheel' and excludes the implicit assumption of superseded conditions.

### 9.2.2   Using the PROPER model for discovery of CRC® elements

Conceptual design according to the C2C® philosophy will always be an iterative growth process, rather than a stacking process, because a multidisciplinary approach is needed in finding the C2C® elements. According to Veeke et al. (2008), a growth process is iterative throughout the whole structure, and in this process we are repeatedly faced with the task of developing new alternatives, choosing from these and testing that choice. It is crucial to a growth process that the design grows through each choice, but we must also solve the problems that are caused by these choices. Partial problems must be isolated, shared between the involved disciplines and solved separately.

Stagnation occurs regularly in this iterative process. We must return to a previous step because it appears that we cannot solve the problems caused by the last but one choice, and so on. According to the PROPER model, maturation periods are continuously required in order to progress, and this is also the case in the innovation process.

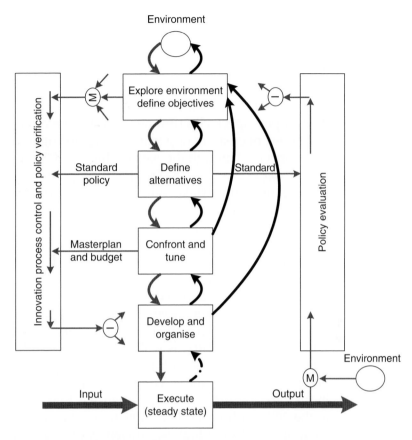

**Figure 9.4   The innovation model: function model for innovation in an aspect system.** *Source*: Veeke et al. (2008). Reproduced with kind permission from Springer Science + Business Media B.V.

Figure 9.4 illustrates this process, in which the iterative actions for solving the partial problems are indicated by curved arrows. Designing according to the C2C® philosophy inevitably complicates the process enormously because of the demanded integration of environmental requirements into product design, production, commercialisation, (re)use and consumption (Vezzoli et al., 2010). C2C® brings together huge quantities of data and relations with stakeholders from different backgrounds; to work together from different points of view and to make any necessary adaptations to the tools for the task, requires professionals with special knowledge. Ideally they work for a company with an interdisciplinary profile which gives them the skills to guide this more complicated design process. This is not just about product or process innovation in relation to sustainability but it goes beyond; as Braungart says, 'focus on the triple top line – beneficial for the environment, lucrative for the economy and good for society'.

## 9.3    Cradle to Cradle for the dairy industry

### 9.3.1    Starting point

From a historical perspective, the dairy industry already has on the 'proper' spectacles (Braungart & McDonough, 2008) for the right view of creating opportunities according to the C2C® business concept. Through the partnership of all farmers in a certain region, *cooperatives* such as FrieslandCampina in the Netherlands were set up to undertake milk processing, thus safeguarding the processing, distribution and marketing of the farmers' dairy products. In addition, product and technology development, export promotion and quality certification were often embedded in a cooperative structure.

So thinking *in the chain* and awareness of social and environmental issues have been part of dairy policy for a very long time; effective partnering as a key success factor in the C2C® business concept has already long been embedded in the dairy industry too.

In the dairy industry as well as in the beer industry, the optimisation of processes with respect to energy and waste reduction is already well ahead, compared with other food processing industries. An important consequence is that the dairy industry in the 20th century has been able to evolve all its processes and technologies to the highest efficiency, of course also thanks to the fact that milk, like beer, is in a liquid state during the longest part of most processing lines (so pumping, heat exchange and filtration were relatively easy to implement and evolve).

In the dairy industry this means a growing awareness of being at the end of the road to higher (eco-)efficiency; the reuse of valuable materials by separation and reuse of recyclable materials in lower applications are more and more seen as 'dead ends' (Vezzoli et al., 2010) with downcycling on all fronts seen as a result that is no longer acceptable to retailers and consumers. Further, the dairy industry understands that modern consumers are well aware that cows are responsible for a large part of greenhouse gas emissions (methane) and that this fact may be damaging the 'green image' of the dairy industry. At the beginning of the 21st century one can conclude that supply chain awareness has been historically embedded in the dairy industry for over a century and thus offers this industry a solid base for a switch now from a strategy of eco-efficiency to the strategy of eco-effectiveness.

This all means that the dairy industry is still in the vanguard, compared with other industries, for 'giving a positive agenda the central role in the design and manufacture of products and services, in which the synergy between economic (business), ecological and social objectives can be strongly promoted' (Vezzoli et al., 2010). Or, as stated by Rossy et al., 'By starting to engage the C2C® business concept in a growth path to be developed to newly set goals, the resulting continuous improvement to eco-effectiveness' can keep the dairy industry in this vanguard position in industry (Rossy et al., 2010).

In summary, the dairy industry is still in a vanguard position, in choosing to implement a sustainable approach according to the C2C® principles:

- Dairy cooperatives have a long history in the food industry.

- Dairy still has a strong 'green image' for consumers.

- Dairy has an existing awareness of the mutual dependence of stakeholders along the whole chain.

- The need for change to effectiveness instead of improving efficiency is not up for discussion in the dairy industry, on the assumption that the industry has already achieved a high level of efficiency in all dairy processing equipment.

- Dairy farmers have always been aware of their role as a human link between nature and nurture, and therefore by tradition accept responsibility for sustainability.

- Farmers have always lived (and still do!) in a sustainable way in their local environment, thus maintaining the character of the ecological, social and economic systems of their home village.

- New dairy products are finding a place in existing markets and also entering new markets (e.g. the soft drinks market) and are able to make use of the necessary 'green and healthy image' as a unique selling point (USP).

Through effective partnering between farmers, the dairy industry, retail and other stakeholders, many C2C® elements for improvement of sustainability are within reach, when mutual dependability in the total chain can be restored within a new modernised structure.

### 9.3.2   C2C® applied to the dairy chain

From the process perspective, milk is of course a raw material at the start of the chain for all kinds of dairy products and has very specific qualities, most of these directly related to the source, i.e. the cow and her feed quality.

At the end of the chain the identity and specifications of consumer goods of dairy origin can differ enormously: from plain (pasteurised) white milk in pouches or bags and packs to all kinds of drinks, desserts, cheese types, food ingredients, meat substitutes and even packaging materials (e.g. polylactic acids or PLAs).

In the search for innovations using C2C® principles, the start should be to take into account the destination of the milk at the end of the chain and then carry out a conceptual analysis of the related process flow, according to the first C2C® principle, 'waste equals food'.

In such a process flow diagram, the incoming and outgoing mass flows can be separately identified, as well as their specifications; this overview makes clear where materials in the diagram tend to flow to degradation or, even worse, tend to become waste via this specific partial flow. These discovered downgrades should now inspire a search for upcycling opportunities, which in most situations is possible via (new) stakeholders. These new stakeholders have their own disciplines and specialties. It is important to make all participating disciplines and specialists aware that they will now 'start the design from scratch', thus freeing the path for the creation of new opportunities. Nowadays, in rapidly changing markets and with developments in information technology, this multidisciplinary communication is very necessary to do and easy to organise.

The total chain of dairy production processes can functionally be analysed in terms of:

- mass balance (with C2C® spectacles: waste equals food);

- energy balance (with C2C® spectacles: use of solar income);

- people flow;

- information flow (with C2C® spectacles: acceptance of mutual dependence/partnership);

- value additions in the chain (with C2C® spectacles: celebrate diversity).

Analysis (through C2C® spectacles) of these systems throughout the whole chain, on headline numbers as well as on elements within the whole, can reveal opportunities through interactions (exchanges, conversions, recoveries, upcycling, buffering, etc.) or by conceiving other functional solutions.

The discovery of promising innovative interactions between systems requires experience of and broad insight into the total chain, as well as proper interdisciplinary communication between all team members. In almost all cases the feasibility of the conceived C2C® elements becoming real projects appears to be dependent on being able to link relevant (new) stakeholders to the newly created (sub)systems. One could say that 'acceptance of mutual dependability/partnership (as the fourth C2C® principle)' (Braungart & McDonough, 2008) of course comes very close to the principle of 'celebrate diversity'; nevertheless, using it as a separate principle gives a sharper focus through the spectacles of C2C®.

By definition, chains in the dairy industry are long and complex, in the number of links as well as in losses of quality because of handling risks throughout the chain. As a first step it is advisable to gain a sharp view of the total chain, to clearly recognise the influence of all aspects of the chain on the quality items (at the input side of the factory as well as at the output side of the product flow) affecting the performance of the new dairy plant. This will offer more and better opportunities to make the design more effective, when applying the C2C® principles to this view of the dairy chain analysis at the start of plant design.

Looking closely at the chain in this way reveals the important role of the time factor in the total chain. And time always costs money in the food industry. As time passes it almost always shortens the shelf life of products; time creates the need for conditioned storage areas and transport systems; the availability of raw materials compared with fluctuating market demands from the consumer's side are almost never in balance; the influence of the season on the volume of food and on milk productivity is often contrary to climate-induced consumption patterns; the availability of 'free energy sources' (e.g. solar or waste heat) are difficult to synchronise fully with energy demand profiles of the processes; and so on.

So 'time' should be approached as one of the systems to be analysed across the total chain. And, as we all know, time is scarce in our modern world, so time is always a main issue in the design, for the design process itself and for the primary goals of the plant or process to be realised.

Obviously the reflection of the dairy chain is visualised best by the materials flow (C2C®: waste equals food) from cow to consumer, but one must be careful to involve energy (C2C®: use of solar income) and value addition (C2C®: celebrate diversity) too, when analysing the chain of the plant to be designed. This is significant not only for success in the revelation of innovative opportunities by interactions, but also for the creative conception of other functional solutions.

To aid in generating the total overview of the dairy chain for the plant to be designed, it is advisable to split up the chain into at least three parts for easier discovery of all stakeholders and the hidden opportunities for innovation. These are:

- the pre-plant part;

- the production plant itself;

- the post-plant part.

With the 'C2C® spectacles on the heads of the multidisciplinary design team members, all of them focused on the collectively defined C2C® ambitions' and this generated overview of the total chain, now lying on the table in three parts, the creative process in search of C2C® elements can make a start.

Besides experience in C2C® design processes, the specialist craftsmanship and knowledge of the participating team members, and the way the project team is organised and structured, a proper use of all type of design support tools (Six Sigma, LEED, REACH, LM, Trias Energetica, etc.) by the members are important for achieving good results.

Some of these tools are monodisciplinary, some of them support interdisciplinary communication, some of these are of temporary value ('fashion'), and some are no longer valid or do not interact with other systems. For all the tools, two questions apply in this design process:

(1) Do the tools contribute by improving analysis and overview in the complexity of the total C2C® design for all team members?

(2) Do the tools support all members to understand 'each other's language' better when the tools are applied?

It's all about communication and creative thinking: innovations are created by using diverging techniques (Byttebier & Vullings, 2007) of which there are several ('creative perception', 'postponing judgement', 'flexible association', 'dissociation/pattern breaking', 'resociation/linking back', etc.), but 'they all need to be fed with trustworthy feed'.

So always pay attention at the start of the conceptual design to make *all members* confident with clear (but not detailed) mass and energy balances, related to the separate black boxes in all phases, with clearly positioned connection points between the boxes, and make all members aware of the information flow as well as the people flow along the chain. The importance of the aspects 'time' and 'environment/society' should be explained as well.

During the design maturation period, important changes to the designed systems will be introduced. When this occurs, review sessions for the whole team should be organised as quickly as possible, to bring together the thinking of all members to the top layer in the design. From there on they can diverge again, but now with augmented understanding of the design hierarchical points, which applies for all disciplines and all members. When the design is properly reviewed, interdisciplinary exchanges and communication will take place in an effective and timely manner. As a result, the necessity for frequent reviews will be reduced and the risk of hidden mistakes in the monodisciplinary parts of the design will be reduced as well. Multidisciplinary design, when enthusiastically and professionally organised in this way, will improve the commitment and focus of all members and create 'a coherent design team' community.

### 9.3.3   *A case study*

An interesting example of a new approach in the dairy industry, inspired by the C2C® principles, is the *DairyLink* concept, a cooperative design project between VMEngineering BV and Royal Haskoning. In this concept the fresh milk production for a region will be produced in a cluster of smallish dairy plants; much smaller than the existing large plants, whose design was driven more by efficiency than by effectiveness.

The design objectives for each small DairyLink satellite in this new cluster of fresh milk plants are:

- capacity per satellite approx. 50 million litres milk per annum (150,000 L/day);

- farmers supply milk and manure;

- farmers are within a range of approx. 40 km;

- local transport powered by biogas;

- 100% climate neutral: no emission of greenhouse gases, no waste water, zero overall energy usage;

- packaging materials climate neutral and recovered by own digesters;
- partnerships with retail and catering industry on collecting organic waste for digesters;
- lower number of full-time employees per shift per satellite, therefore necessity of high level of automation of production and quality control;
- extremely modular (processes, packaging, the building and services) for fast and easy implementation;
- operational and technical key performance indicators at least comparable with larger plants;
- algae for wastewater treatment and feed production;
- photovoltaic solar panels (on plant roofs and farm roofs);
- selection of building structure and materials based on C2C® principles.

The DairyLink concept fits amazingly well with current developments in society towards local, regional, 'zero km' products, Slow Food, etc. (Vezzoli et al., 2010).

It is likely that at the end of 2012 two European dairy companies will implement the DairyLink concept in their strategies.

Other great opportunities for C2C® in dairy are the recovery of water from all drying processes; milk of course consists almost entirely of water (of controlled 'food grade' quality) when it comes in and this water is nowadays still expelled as wastewater from nearly all drying plants.

One could say that the cow has invested a great deal of energy to take in all the water, and then the milk is processed in a very energy-consuming way to get the water out of the milk and emit it as wastewater! Analysing this chain in another way would take into account the advantages for local people in specific (arid) regions, if (food grade) water was to be effectively recovered; this really would offer a better incentive to prevent the production of 'wastewater' from the milk flow.

## 9.4 Conclusion

Starting the design of a new dairy plant by first analysing the total value chain of the milk 'through the spectacles of C2C®', in order to set the design goals, offers additional opportunities for the discovery of C2C® elements as very effective innovations, which as separately implemented projects will really contribute to business results as well as to the realisation of environmentally positive operational aspirations.

Of course, the resulting greater complexity of the total design approach needs to be handled in a professional way by an experienced multidisciplinary design team, that makes use of sophisticated methods and tools for interdisciplinary design communication and an efficient iterative decision-making process, thus translating the design challenges from the start into an inspired, effective and sustainable total design of the plant as a result.

For this systems approach we recommend the PROPER model (Veeke et al., 2008), specifically because it offers a very good frame of reference for logistical systems such as the value chain of milk.

We believe the highly developed dairy industry of the 21st century has now gained the necessary momentum to create designs that are inspired by the C2C® principles for achieving innovative, sustainable and effective new solutions in milk processing, within a roadmap that will contribute to the successful further embedding of the dairy industry in the society of the future. The historical relationship of dairy with nature ('biodiversity') and consumers ('nutrient') has always been very strong, and C2C® can only add more value to the very strong image that dairy has already established in society over many years. Furthermore, the cooperative structure ('partnership'), that until now has characterised the dairy industry for several decades in a number of regions, without doubt gives witness to the successful implementation of the three C2C® principles in this industry.

The image of dairy has always been 'green', but the human interconnectedness of dairy with nature could easily be strengthened by implementing the C2C® principles from the start as inspiration for a new integral design approach that socially, economically and ecologically will fit in with the future demands for a bio-based economy in a robust and sustainable way.

The two most important stakeholders in the contemporary value chain for milk are already focused on sustainability:

1.  At the beginning of the chain, farmers have historically had this focus.

2.  At the end of the chain, consumers over recent years have made growing demands for sustainability in the products they consume.

In connecting to those consumer demands, the dairy industry has already made a start with sustainable design, and the 'spectacles of C2C® principles' can help to improve and speed up the focusing. This applies to materials, energy, information, people and time.

And as most dairy products remain perishable after processing, extra emphasis will appear justified in tailoring human systems and industries to local economies in order to achieve optimal sustainability in dairy design, as affirmed by Braungart and McDonough in *Cradle to Cradle: Waste = Food*.

## References

Braungart, M., McDonough, W. (2008) *Cradle to Cradle: afval = voedsel*. Heeswijk: Search Knowledge BV.

Byttebier, I., Vullings, R. (2007) *Creativity today: tools for a creative attitude*. Amsterdam: BIS Publishers.

Rossy, A., Jones, P.T., Geysen, D., Bienge, K. (2010) *Sustainable materials management for Europe: from efficiency to effectiveness*. Informal Environmental Council.

Veeke, H.P.M., Ottjes, J.A., Lodewijks, G. (2008) *The Delft Systems Approach: analysis and design of industrial systems*. Springer.

Vezzoli, C., Orbetegli, L., Cortesi, S. (2010) *C2C Perspective study: Industry* [online]. www.c2cn.eu.

# 10

# The future of sustainable dairy production

Peter de Jong

NIZO Food Research BV, Ede, The Netherlands

**Abstract:** In this chapter some strategies are discussed for coping with future issues for the dairy sector. Two parallel approaches are suggested: chain innovations with breakthrough technologies; and communication to stakeholders. The dairy industry is an intrinsically sustainable food sector delivering a great deal of nutrients relative to its impact on climate. To get this message across, a considerable communication effort is needed.

**Keywords:** breakthrough technologies, business case, chain innovation, communication, future

## 10.1  Future relevance of sustainable dairy

It is very reasonable to expect that the global population will continue to increase. As a consequence there will be a worsening shortage of resources to feed the billions of people (Jansa et al.,

*Sustainable Dairy Production*, First Edition. Edited by Peter de Jong.
© 2013 John Wiley & Sons, Ltd. Published 2013 by John Wiley & Sons, Ltd.

2010).Food security in 2050 will not be easy to achieve. There are challenges to be solved. Therefore it is clear that sustainability must be on the agenda of every dairy company. It should be clear that greenhouse gas emissions in terms of $CO_2$-eq per kg product are only one component of the total environmental impact. The social and economic impacts of dairy production need also to be considered.

Apart from day-to-day optimisation, two parallel approaches can be distinguished: chain innovations and communication to stakeholders. Especially for innovations in the production chain, economic feasibility is an important prerequisite. In this final chapter the main conclusions of the previous chapters are transformed into concrete road maps for the near future.

## 10.2   Next steps in chain innovation

### 10.2.1   *Measuring sustainability*

The dairy industry does not see labelling of carbon footprints as a good idea (IDF, 2008). However, large players in retail have already started schemes to move toward a situation in which all suppliers will need to provide information on impacts or footprints of their products. Forced by public opinion and retail, the industry will increasingly use life cycle analysis (LCA) as an instrument for business decisions and customer communication. Although a number of companies give an indication of the carbon and even the water footprint of their products, these numbers cannot be used for benchmarking of different food products. The main challenge here is to find a standardised guideline and calculation method for determination of the carbon footprint of food products. Allocation problems have to be solved and sound databases of LCA data should be available to everyone.

Until now, the measurement of sustainability in the dairy industry has concentrated on the carbon footprint of dairy products. However, sustainability is much more than that. Water footprints, acidification, eutrophication and land use also have their impact on sustainability. In the future, algorithms have to be found that combine all these factors together with the carbon footprint to give a'sustainability indicator' in a transparent way.

### 10.2.2    *Key role of the farm*

The major part of the carbon footprint of dairy products is determined in the production chain before the processing of raw milk into dairy products. An FAO report (Gerber et al., 2010) stated that, globally, cradle to farm gate emissions contribute 93% of total $CO_2$-eq emissions. Methane, originating mainly from cow metabolism, contributes 52%. This means that a lot of effort has to be made to reduce this and to improve farm management in general. The following roads to solutions may be considered:

- cow breeding not focused on production quantities but focused on carbon footprint and other sustainability factors (e.g. water footprint, acidification, land use);
- feed optimisation (e.g. type of feed, source of feed, manufacturing, composition);
- decrease of methane emission by manipulation of the cow's microflora and digestion;
- extraction of methane from the air by advanced separation technologies;
- application of renewable sources (i.e. energy and water) and biogas;
- farm management (geographical location, cattle size, outdoor grazing).

It is clear that the technical and economic feasibility of these solutions will vary. However, an important criterion is: what is the resilience and sustainability of the farm itself? Whereas in Europe a significant proportion of farms achieves a positive profit, there also is a significant proportion earning a negative economic profit. Implementation of new technologies and making new investments will not be feasible for the latter group. This implies that arrangements need to be put in place so that the farmers receive economic benefits from being sustainable.

### 10.2.3    *Development of breakthrough technologies*

Optimisation of production lines with respect to costs and sustainability should always be on the agenda of a plant manager. It will lead to annual savings in the order of 10% to 20%. However, to make a significant step in reducing carbon footprints, new (break-through) technologies need to be developed and implemented. The following examples may even be feasible in the midterm:

- Fully automated, unmanned production units of dairy products. Eventually, these units can be placed near large farms. Challenges: model-based process control, quality control.

- Alternative unit operation for drying. For example, high solids concentrates instead of powder, vacuum drum drying. Challenges: product quality and functionality.

- Waste valorisation by advanced separation and purification technologies. Challenges: product functionality, return on investment.

- Self-cleaning equipment by using ultrasound technologies. Challenges: return on investment.

### 10.2.4 Packaging

The carbon footprint of packaging is of the same order of magnitude as that of dairy processing in factories (IDF, 2009). For more or less the same reasons as the dairy industry, the packaging industry has sustainability high on its agenda. The following solution directions can be identified to improve the primary packaging of dairy products:

- end-of-life treatment (recycling);

- material use, green sourcing (biodegradable, renewable materials and manufacturing);

- packaging size.

Packaging has a big impact on dairy products. Among other things it influences the attractiveness, convenience, shelflife and, last but not least, the safety of the product. As a consequence manufacturers of packaging need to cooperate with dairy companies to achieve new sustainable packaging solutions.

### 10.2.5 Cradle to Cradle

In order to, for example, halve the carbon footprint of milk it is necessary to fully revisit the production chain and implement breakthrough technologies which minimise energy use and product loss. It can be effective to apply the Cradle to Cradle concept, meaning roughly that no waste is generated (waste = food) and production is energy neutral. This would also mean that the carbon footprint goes down to almost zero. Although one can debate the feasibility of this for the near future, it stimulates the generation of new production concepts in which sustainability is the main driver. This may result in new insights with economic benefits as well.

## 10.3  Communication and marketing

As described in Chapter 1, communication and marketing of sustainability is a key factor for the future viability of the dairy sector. In 2011 the Global Dairy Platform started an initiative communicating that the dairy sector needs to be recognised as a vital source of essential nutrients and distinguished for proactively managing its relationship with the environment. Moreover, the Platform recognised that there is an urgent need to revive consumer confidence in dairy products as being healthy, natural, enjoyable and tasty.

When we discuss global food security for the future it is not appropriate to benchmark on carbon footprint alone. To feed a growing population we need nutrients and not food products with a low footprint as such. It would be worthwhile to promote benchmarking of dairy products based on parameters such as nutrient density related to climate change.

## 10.4  Business case: people, planet and profit

Since sustainability has to do with people and their environment, a sustainable company cannot be focused on profit and shareholder value alone. As Tachi Kiuchi, former CEO of Mitsubishi Electronics, said: 'The whole essence of business should be responsibility ... we don't run companies to earn profits. We earn profits to run companies' (Covey & Merrill, 2006).

*For the dairy industry this means that they need to earn profit in order to feed the people in a healthy environment for the coming decades and centuries.*

From that perspective, profit is still an essential building block of sustainable dairy and should always be on the agenda (see

**Table 10.1** Example of a business case format for implementing new technologies in a sustainability-focused dairy company.

| Business case<br>Objective (70% decrease in product loss by implementing new technology increasing shelf-life) | People | Planet | Profit |
|---|:---:|:---:|:---:|
| **Drivers** | | | |
| + Strategic goals (be number one in sustainable dairy production, climate neutral production, …) | ☺ | ☺ | ☺ |
| + Context (consumer organisation focus on raw material efficiency, food scarcity, financial crisis, …) | ☺ | ☺ | ☺ |
| **Benefits** | | | |
| + Production costs reduction (euro/kg): … | | | ☺ |
| + Carbon footprint reduction ($CO_2$-eq./kg): … | | ☺ | |
| + Water footprint reduction (kg water/kg): … | | ☺ | |
| + Improvement of nutrient density to climate impact index (NDCI): … | | ☺ | |
| + Specific benefits (increased capacity of equipment (h/unit), less spoiled food in consumers' refrigerator (unit/consumer), …) | ☺ | ☺ | ☺ |
| **Risks** | | | |
| + People (increased complexity for operators, …) | ☺ | | |
| + Product quality (increased probability of contamination, decreased taste and texture after 3 months, …) | ☺ | | ☺ |
| + … | | | |
| **Costs and planning** | | | |
| + Cost of development and implementation (euro): … | | | ☺ |
| + Cost breakdown and planning of activities: … | | | ☺ |
| + … | | | |
| **Investment** | | | |
| + Investment (euro): … | | | ☺ |
| + ROI (%): … | | | ☺ |
| + … | | | |

also Chapter 1). In the end, without profit there will be no (sustainable) dairy production at all.

On the other hand, given the clear demographic trends, in the long term without sustainability there will be no profitable company either. As a consequence, future business cases for technological changes in dairy production may look as shown in Table 10.1.

## 10.5  Conclusion

The dairy industry is an intrinsically sustainable food sector delivering a great deal of nutrients relative to its impact on climate. To get this message across a considerable communication effort is needed. Meanwhile the dairy sector can strengthen its position on sustainability by promoting a standardised measure of sustainability for food which is not just based on the carbon or water footprint. Therefore it is necessary to develop measures and technologies reducing the environmental impact of the farm, dairy processing and dairy packaging. In parallel, revisiting the dairy production chain by using the Cradle to Cradle concept may give new insights for an even more sustainable future in the long term. Recent business cases of dairy companies show that they are ready for the next challenges.

## References

Covey, S.M.R., Merrill, R.R. (2006) *The speed of trust: the one thing that changes everything*. New York: Simon & Schuster.

Gerber, P., Vellinga, T., Opio, C., Henderson, B., Steinfeld, H. (2010) *Greenhouse gas emissions from the dairy sector: a life cycle assessment*. FAO report. Rome: Food and Agriculture Organization.

Global Dairy Platform (2011) www.globaldairyplatform.com/Documents/GDP%202011%20Membership%20Brochure.pdf.

International Dairy Federation (2008) Consideration of future IDF work programme with regard to sustainability with focus on environmental/ecological aspects. Questionnaire 0608. Brussels: IDF.

International Dairy Federation (2009) Enviromental/ecological impact of the dairy sector: literature review on dairy products for an inventory of key issues list of environmental initiatives and influences on the dairy sector. *Bulletin of the International Dairy Federation*, no. 436. Brussels: IDF.

Jansa, J., Frossard, E., Stamp, P., Kreuzer, M., Scholz, R.W. (2010) Future food production as interplay of natural resources, technology and human society: a problem yet to solve. *Journal of IndustrialEcology*, 14: 874–877.

# Index

Note: Page numbers in *italics* refer to figures, those in **bold** refer to tables and boxes and those with suffix 'n' refer to footnotes